Nuclear Waste

Management, disposal and governance

Nuclear Waste

Management, disposal and governance

Edited by
Klaus-Jürgen Röhlig
Technische Universität Clausthal, Institute of Disposal Research, Clausthal-Zellerfeld, Germany

IOP Publishing, Bristol, UK

ISBN 978-0-7503-3095-4 (ebook)
ISBN 978-0-7503-3093-0 (print)
ISBN 978-0-7503-3096-1 (myPrint)
ISBN 978-0-7503-3094-7 (mobi)

DOI 10.1088/978-0-7503-3095-4

Version: 20220401

IOP ebooks

British Library Cataloguing-in-Publication Data: A catalogue record for this book is available from the British Library.

Published by IOP Publishing, wholly owned by The Institute of Physics, London

IOP Publishing, Temple Circus, Temple Way, Bristol, BS1 6HG, UK

US Office: IOP Publishing, Inc., 190 North Independence Mall West, Suite 601, Philadelphia, PA 19106, USA

Cover image: Reproduced courtesy of SFOE

To those seeking solutions.

Contents

Introduction

How should we deal with the radioactive materials and substances produced by the operations of nuclear power plants or by other activities involving radioactivity? The latter includes practises that use, and benefit from, radiation and radioactivity (e.g. radiodiagnostics and radiotherapy or materials analysis in industry), but also cases in which radioactive concentrations just 'happen' in materials[1] or objects as a byproduct of activities whose objectives have nothing to do with radioactivity or nuclear fission, examples being coal incineration, the oil and gas industry, and other technologies in which materials are extracted from underground (e.g. fracking, geothermy). Although all these areas are important, it is probably fair to say that, due to the high activity, (radio-)toxicity, and longevity of the materials produced, the issues of nuclear power and of nuclear weapons in particular have created awareness of, and controversies about, the management of such materials. Actors far beyond the relatively small community of specialists who are legally, administratively, and technically responsible for this management have became concerned and involved. Debates, sometimes evolving into serious conflicts, have taken and are taking place in several countries—often, but not only, around the siting of disposal facilities.

For spent nuclear fuel (SNF) from power plants, a material with a particularly high radioactivity content and an associated high and long-lasting hazard potential, the question of whether it should be considered as waste or as a resource which can be recycled (since it still contains considerable amounts of fissile materials) is being debated—the answers vary from country to country and also over time. But even if the SNF is recycled (reprocessed), residues with high levels of radioactivity (and by-products with lower levels) arise, which must be treated as waste and managed safely in accordance with their hazard potential.

To the editor, after decades of working in the field, the topic of this book—the management of such nuclear or radioactive waste[2]—is still challenging and highly attractive at the same time. There are two reasons for this: first, there are the societal causes for the existence of the waste and the controversies mentioned above which make the issue a technical, societal, and political one—with all the insights, experiences, contacts, and debates which are very interesting for a scientist.

The second reason has much to do with the objectives of this book: Nuclear waste management involves numerous—and very different—academic disciplines. Nuclear physics, nuclear chemistry, nuclear engineering, and radiation protection immediately come to mind, but the importance of others, such as materials science, civil engineering, the geosciences, and mathematics is also rather obvious—as is the relevance of law, sociology, history, the political sciences, human geography, ethics, and many other disciplines from the humanities and social sciences. These disciplines cannot address the issue appropriately if they work in isolation—

[1] Often referred to as 'naturally occurring radioactive materials' (NORM) and 'technologically enhanced naturally occurring radioactive materials' (TENORM).

[2] We use the two terms 'radioactive waste' and 'nuclear waste' synonymously, as preferences concerning this terminology vary amongst academic disciplines.

interdisciplinary communication, interaction, and co-operation are indispensable. And although interdisciplinarity is a fashionable term and therefore often used superficially these days, those who have to implement and to live it know that it is attractive, interesting, and challenging at the same time.

In more practical terms: people working in the field of nuclear waste management need to be highly qualified specialists who also have good overview and context knowledge. Ideally, this should be supplemented by good communication skills. The latter are important when communicating with other specialists in order to obtain sound results, but also necessary to communicate with other individuals, e.g. local stakeholders, members of the interested public, journalists, etc. Such a need or occasion for 'external communication' might even come as surprise to people working in the field—it is not always part of an elaborated and well-planned communication strategy. And good overview and context knowledge is essential on such occasions as well.

This book aims to provide such knowledge. Its target audience has a scientific background, it is consisting of students, beginners in the field and specialists working in nuclear waste management who want to understand the overall picture and to get an overview of the work being done by other specialists. Scientists working in other fields might also find it interesting. 'Selective reading' is encouraged, as the authors aimed to write chapters which can be read on their own.

Naturally, this type of book has its limits: details need to be looked up elsewhere, but we (the authors and the editor) have high hopes that we can raise awareness about the central issues and interrelations and encourage readers to look for the important things in the right places. Numerous references are included for further study. Also, not all the issues related to nuclear waste management are covered—important (but intentional) omissions include, for example, waste from military activities. Other topics, such as the economics of nuclear waste management and human factors in safety-relevant systems might be addressed in future publications. Finally: we did our very best to provide up-to-date knowledge, but we address an evolving field. What the reader will find in this book is the current state of knowledge.

As mentioned above, nuclear waste management is an undertaking that involves interactions between scientific, technical, and societal aspects, none of which can be ignored—it can be described as a 'socio-technical' issue. For orientation purposes, we decided to write a book with a first part focussing on natural sciences and technology and a second focusing on societal aspects—in the knowledge that there is no real dividing line between the two. We start by providing basic knowledge about the features which make our topic unique: radioactivity and ionising radiation, their origins and nature, the associated threats and risks for human health, and radiation protection as means for addressing such risks (part I, chapter 1 by Clemens Walther). We then go beyond questions of pure 'waste management': as we mentioned above, the fate of the highly radioactive SNF is not necessarily to become, or to be declared, waste, but the decision about the path the SNF will take will have a significant impact on nuclear waste management. Therefore, we sketch and assess considerations and possibilities 'from cradle to grave' for the whole so-called 'fuel cycle' (which is not a cycle in the strict sense of the word) and related reactor technologies (chapter 2, Wolfgang Kröger and Andrej Stankovski). This is all the more important, since, in debates about nuclear waste, arguments about reducing or even nullifying the hazard caused by radioactivity often arise—here, we describe the only options for doing so which are actually considered and perhaps realistic from a technological point of view.

As we said above, there are many waste origins other than nuclear power generation (medicine, industry, research, etc.), for which the ensuing chapter (chapter 3, Michael Ojovan) provides an overview. When a society knows about the kinds and quantities of nuclear waste it has to deal with, it can make up its mind about the ways to do so: a strategy has to be developed and implemented that accounts for high-level objectives, such as safety, security, and the protection of humans and the environment; ideally, it should anticipate surprises and potential obstacles (chapter 4, Klaus-Jürgen Röhlig). An essential part of such a strategy will be the production of waste forms suitable for the actions to be undertaken (such as transport, storage, and disposal, see below), the so-called 'conditioning' (chapter 5, Michael Ojovan). It will be necessary to hold waste materials temporarily in dedicated facilities ('storage') as well as to transport them from location A to location B (chapter 6, Maik Stuke, Mathias Häring, and Jörn Becker).

Ultimately, there should be safe, secure, and durable solutions for the materials ('strategy endpoints', chapter 7 by Klaus-Jürgen Röhlig). Various such options have been discussed in the past, including launching them into outer space or emplacing them into subduction zones or ice shields, or disposal at sea (which has been done in the past for some wastes). The present state of the art and the relevant regulations require the materials to be concentrated and contained in dedicated facilities ('repositories') without the intention of retrieval ('disposal')[3]. According to the so-called 'graded approach' advocated by the International Atomic Energy Agency (IAEA), the strength and sophistication of the engineering of such facilities has to increase along with the hazard potential of the specific waste type to be disposed of. Moreover, increasing hazard should go along with an increase in the depth at which the facilities are to be emplaced in the ground. For the materials with the highest radioactivity contents and lifetimes, so-called 'geologic' or 'deep' disposal, i.e. disposal at depths of several hundred meters, is advised. Depending on the nature of the geologic formation used (the 'host rock'), different technical concepts will be applied (chapter 8, Klaus-Jürgen Röhlig). In any case, safety is a decisive issue when decisions are to be made about the siting, development, construction, operation and closure of such facilities—a so-called 'safety case' is therefore an important tool that informs decisions in a disposal programme (chapter 9, Klaus-Jürgen Röhlig).

This 'safety case' can be seen as one of several elements that connect the technical and societal aspects of nuclear waste management. Legislation in the form of acts and ordinances, their development, implementation, and application are another connecting element between the societal and the technical fields (part II, chapter 10

[3] This is not to say that retrieval is impossible, and, in some cases, an option for retrieval ('retrievability') might be legally required and technically implemented.

by Hans Wanner). Safety is an important issue for legislation, and, at the same time, a central concern for many interested individuals and groups. Therefore, it is important to know how such basic concepts as safety, or its counterpart, risk, are understood and perceived by individuals and groups—these understandings and perceptions in many ways form the approaches these individuals take in discussions and actions (chapter 11, Anne Eckhardt). Perceptions, attitudes, and therefore discussions and actions, have their roots in history—as does nuclear waste itself. After all, it is a human-made product, and thus there are social, political, and cultural reasons why it exists in the first place. Actors in the field of nuclear waste management will benefit from knowledge about its history and the role politics and politization have played in this story (chapter 12, Karena Kalmbach). Such (and other) knowledge is needed when it comes to the question of how to involve various and diverse stakeholders in the development of nuclear waste management programmes and projects (chapter 13, David Brazier, Jo-Ann Facella, Pascale Jana Künzi, and Erik Setzman). Ethical considerations, e.g. about safety and fairness will play an important role in any discourse about nuclear waste management, but the discourse itself should also be assessed from an ethical point of view (chapter 14, Konrad Ott). All this plays a particular role when it comes to choosing sites for repositories, but the geologic boundary conditions are also of the utmost importance. In other words, this is another point of connection between science, technology, and society (chapter 15, Klaus-Jürgen Röhlig). This book concludes with a recent example of a site selection procedure in Germany, which has just begun and aims to integrate all these elements (chapter 16, Armin Grunwald).

For the editor, one of the most attractive aspects of interdisciplinary work in nuclear waste management is the opportunity to become acquainted with clever and interesting individuals with widely varying approaches and views. It is not always easy but always rewarding to arrive at a common language and understanding—lifelong learning at its best. Fruitful professional relationships form, which sometimes evolve into friendships. The international and interdisciplinary authors' team came together on this basis. It was a privilege and a pleasure to work with this team, to discuss and to learn from each other. I would like to thank all of its members for their continued engagement, effort, and intensive work. I hope that readers will perceive some of the fruitful exchanges we had, and that they enjoy reading and learning by means of this book.

Klaus-Jürgen Röhlig, editor
Clausthal-Zellerfeld, Germany, September 2021

Editor biography

Klaus-Jürgen Röhlig

Klaus-Jürgen Röhlig, born in 1958 in Leipzig (Saxony, Germany) is a mathematician who received his PhD in mathematical analysis from the Mining Academy (Bergakademie), Freiberg (Saxony) in 1989. His work at the Institute of Energetics (IfE), Leipzig on code development and the numerical simulation of fluid flow and contaminant migration was a first encounter with nuclear waste disposal issues. From 1991 to 2007 he worked at the Gesellschaft für Anlagen und Reaktorsicherheit (GRS) in Cologne (North Rhine-Westphalia). His work included research and development in the areas of numerical simulation, probabilistic methods, and assessment methodology in the context of safety assesments for nuclear waste disposal, the provision of scientific and regulatory advice to the German Federal Ministry for Environment, Nature Conservation and Nuclear Safety (BMU), as well as a temporary secondment to the Organisation for Economic Co-operation and Development (OECD) Nuclear Energy Agency (NEA) in 2000.

Since 2007, he has been a university professor (repository systems) at Clausthal University of Technology (Lower Saxony). His research and lecture interests include safety and performance assessment, nuclear waste management strategies, and disposal concepts. Over the last 10 years, he has become increasingly involved in socio-technical research into nuclear waste management, most notably as Speaker and coordinator of the co-operative projects 'Disposal Options for Radioactive Residues: Interdisciplinary Analyses and Development of Evaluation Principles' (ENTRIA, 2013–2018) and 'Transdisciplinary Research on the Management of High-Level Radioactive Waste in Germany' (TRANSENS, 2019–2024).

He is the chair of the German Association of Disposal Research (Deutsche Arbeitsgemeinschaft Endlagerforschung (DAEF)) and a member of several scientific and advisory bodies, including the Nuclear Waste Management Commission (ESK, advisory body to the BMU, deputy chair 2008–2010), the Integration Group for the Safety Case (IGSC) at OECD/NEA (chair 2010–2015), and the Cantons' Expert Group Safety (KES, Switzerland). In the past, he has provided advice to safety authorities in France, Sweden, and Switzerland.

List of abbreviations/acronyms

ADR	Accord relatif au transport international des marchandises Dangereuses par Route
ADS	Accelerator-driven system
AFR	Away-from-reactor (storage facilities)
AGR	Advanced gas-cooled reactor
ALARA	As low as reasonably achievable
Andra	Agence nationale pour la gestion des déchets radioactifs
APM	Adaptive phased management
ARGONA	Arenas for Risk Governance Project
ARTEMIS	Integrated Review Service for Radioactive Waste and Spent Fuel Management, Decommissioning and Remediation
ASN	Autorité de Sûreté Nucléaire
AVR	Arbeitsgemeinschaft Versuchsreaktor (Jülich), German for: Working Group Experimental Reactor
BASE	Bundesamt für die Sicherheit der nuklearen Entsorgung, German for: Federal Office for the Safety of Nuclear Waste Management
BGE	Bundesgesellschaft für Endlagerung, German for: Federal Company for Radioactive Waste Disposal
BGZ	Gesellschaft für Zwischenlagerung mbH, German for: Federal Company for Interim Storage
BMU	Bundesministerium für Umwelt, Naturschutz und nukleare Sicherheit, German for: Federal Ministry for Environment, Nature Conservation and Nuclear Safety
BSS	Basic safety standards
BWR	Boiling water reactor
Bq	Becquerel (unit of radioactivity, 1/s)
CANDU	Canada deuterium uranium (reactor)
CARL	Citizen-stakeholders, Agencies responsible for radioactive waste management, social science Research organizations, and Licensing and regulatory authorities
CASCAD	Vault storage system (note that this is not an acronym standing for a certain phrase)
CASTOR®	Cask for storage and transport of radioactive material
Cigéo	Centre industriel de stockage géologique
Clab	Centralt mellanlager för använt kärnbränsle, Swedish for: Central Interim Storage Facility for Spent Nuclear Fuel
Clink	Centralt mellanlager och anläggning för inkapsling av använt kärnbränsle, Swedish for: Central interim storage and encapsulation plant for spent nuclear fuel
COWAM	Community waste management
COX	Callovo-Oxfordien (geologic formation in the Paris Basin)
D, D_T	Absorbed dose, dose absorbed by an organ T
DSRS	Disused sealed radioactive sources
E	Equivalent dose
EBS	Engineered barrier system
ECOSOC	Economic and Social Council of the United Nations
EIA	Environmental Impact Assessment

ENTRIA	Entsorgungsoptionen für radioaktive Reststoffe. Interdisziplinäre Analysen und Entwicklung von Bewertungsgrundlagen, German for: Disposal Options for Radioactive Residues: Interdisciplinary Analyses and Development of Evaluation Principles (project)
EOL	End of life
EPR	European Pressurised (water) Reactor / Evolutionary Power Reactor
ERU	Enriched reprocessed uranium
ETH	Eidgenössische Technische Hochschule (Zürich / Lausanne)
EW	Exempt waste
EWN	Entsorgungswerk für Nuklearanlagen, German for: Disposal Company for Nuclear Facilities
FHR	Fluoride-salt-cooled high-temperature reactor
FLiBe	Molten salt made from a mixture of lithium fluoride (LiF) and beryllium fluoride (BeF_2)
FSC	Forum on Stakeholder Confidence
ewG	Einschlusswirksamer Gebirgsbereich, German for: containment-providing rock zone
GCM's	Glass composite materials
GEN-IV	Generation IV (reactors)
GFR	Gas-cooled fast reactor
GHG	Greenhouse gas
GPD	Groupe permanent d'experts pour les déchets
GWd/t HM	Gigawatt-days per metric ton of heavy metal
Gy	Gray (J kg^{-1})
HABOG	Hoog radioaktief Afval Behandelings- en OpslagGebouw, Dutch for: High-level radioactive waste treatment and storage building
H_T	Equivalent dose for an organ T
HLW	High-level (radioactive) waste
HTGR	High-temperature gas-cooled reactor
HTR	High-temperature reactor
HTR-PM	Pebble-bed modular high-temperature reactor
HWR	Heavy water reactor
IAEA	International Atomic Energy Agency
ICAO-TI	Technical Instructions for the Safe Transport of Dangerous Goods by Air
ICRP	International Commission on Radiological Protection
IGSC	Integration Group for the Safety Case (OECD/NEA)
ILW	Intermediate level (radioactive) waste
IMDG-Code	International Maritime Dangerous Goods-Code
INF-Code	International Code for the safe Carriage of Packed Irradiated Nuclear Fuel, Plutonium and High-level Radioactive Wastes on Board Ships
InSOTEC	International Socio-Technical Challenges for Implementing Geological Disposal (EU project)
IRGC	International Risk Governance Council
IRRS	Integrated Regulatory Review Service
IRSN	Institut de Radioprotection et de Sûreté Nucléaire
KBS-3	Kärnbränslesäkerhet, Swedish for: nuclear fuel safety (Swedish disposal concept)

KBS-3H	KBS-3 with horizontal canister emplacement
KBS-3V	KBS-3 with vertical canister emplacement
K_{eff}	Effective neutron multiplication factor
KONVOI	Reactor type (note that this is not an acronym standing for a certain phrase)
LCA	Life cycle assessment/analysis
LFR	Lead-cooled fast reactor
LILW	Low- and intermediate-level (radioactive) waste
LLW	Low-level (radioactive) waste
LSA	Low specific activity material
LWR	Light water reactor
MJ	Megajoule
MVDS	Modular vault dry storage
MoDeRn	Monitoring Developments for safe Repository operation and staged closure (EU project)
Modern2020	Monitoring Developments for safe Repository operation and staged closure 2020 (EU project)
MOX	Mixed oxide (nuclear fuel)
MSR	Molten salt fast reactor
MYRRHA	Multi-purpose Hybrid Research Reactor for High-tech Applications
NBG	Nationales Begleitgremium, German for: National Citizens' Oversight Committee
Nagra	Nationale Genossenschaft für die Lagerung radioaktiver Abfälle, German for: National Cooperative for for the Disposal of Radioactive Waste (Switzerland)
NEA	Nuclear Energy Agency (operating within the framework of the OECD)
NEEDS	New Energy Externalities Developments for Sustainability
NEZ	Nukleares Entsorgungszentrum, German for: Nuclear Waste Management Centre
NIMBY	'Not In My Back Yard'
NFC	Nuclear fuel cycle
NGO	Non-governmental organisation
NORM	naturally occurring radioactive materials
NPP	Nuclear Power Plant
NUHOMS®	NUTECH horizontal modular storage
NWMO	Nuclear Waste Management Organization (Canada)
OBRA	European Observatory for Long-term Governance on Radioactive Waste Management
OECD	Organisation for Economic Co-operation and Development
OHF	Organisational and human factors
ONDRAF/NIRAS	Organisme national des déchets radioactifs et des matières fissiles enrichies / Nationale instelling voor radioactief afval en verrijkte splijtstoffen (Belgium)
OPG	Ontario Power Generation
P&T	Partitioning and transmutation
PET	Positron emission tomography
PSA	Probabilistic safety analysis (assessment)
PUREX	Plutonium–uranium recovery by extraction

PV	Photovoltaics
PWR	Pressurized (light) water reactor
R&D	Research and development
RD&D	Research, development & demonstration
RBMK	Реактор Большой Мощности Канальный (Russian high-power channel-type reactor)
RHWG	Reactor Harmonisation Working Group (at WENRA)
RID	Règlement concernant le transport international ferroviaire de marchandises Dangereuses
RISCOM	Risk Communication initiative
RSC	Rock Suitability Classification system
RSK	Reactor Safety Commission
RWM	Radioactive waste management
RWMC	Radioactive Waste Management Committee (OECD/NEA)
RWMC-RF	Regulators' Forum of the RWMC
RXA	Recrystallized annealed
SCO	Surface contaminated objects
SFOE	Swiss Federal Office of Energy
SFR (part I chapter 2)	Sodium-cooled fast reactor
SFR (part I chapters 4 and 7)	Slutförvaret för kortlivat radioaktivt avfall. Swedish for: Final Repository for Short-Lived Radioactive Waste
SFW	Spent nuclear fuel considered as waste
SKB	Svensk Kärnbränslehantering AB, Swedish for: Swedish Nuclear Fuel and Waste Management Company
SiR	Silicon carbide
SMR	Small modular reactors
SNF	Spent nuclear fuel
SRA	Stressed relieved annealed
SRS	Sealed radioactive sources
SSK	Strahlenschutzkommission, German for: Commission on Radiological Protection
SSM	Strålsäkerhetsmyndigheten, Swedish for: Swedish Radiation Safety Authority
StandAG	Standortauswahlgesetz, Gesetz zur Suche und Auswahl eines Standortes für ein Endlager für hochradioaktive Abfälle, German for: Act on the Search for and Selection of a Site for a Disposal Facility for High-Level Radioactive Waste
STS	Science and technology studies
Sv	Sievert (unit of ionizing radiation dose)
TBP	Tributyl phosphate
TENORM	Technologically enhanced naturally occurring radioactive materials
THTR	Thorium high-temperature nuclear reactor
TRANSENS	Transdisziplinäre Forschung zur Entsorgung hochradioaktiver Abfälle in Deutschland, Transdisciplinary Research on Radioactive Waste Management in Germany (project)
TRISO	Tri-structural isotropic
TRUSTNET	EU Concerted Action on risk governance (note that this is not an acronym standing for a certain phrase)
TWh	Terawatt-hour
URL	Underground research laboratory

VAHA	Vaatimusten hallintajärjestelmän, Finnish for: requirements management system
VALDOR	Values in Decisions On Risk symposium
VHTR	Very-high temperature reactor
VLLW	Very low-level (radioactive) waste
VVER	Водо-водяной энергетический реактор (Russian water cooled and moderated reactor)
VSG	Vorläufige Sicherheitsanalyse Gorleben, German for: Preliminary safety assessment Gorleben
VSLW	Very short-lived (radioactive) waste
WAC	Waste acceptance criteria
WENRA	Western European Nuclear Regulators' Association
WGWD	Working Group on Waste and Decommissioning (at WENRA)
WHO	World Health Organization
WMO	Waste management organization (generic term)
WNA	World Nuclear Association
ZIRA	Zone d'intérêt pour la reconnaissance approfondie, French for: zone of interest for detailed reconnaissance

List of contributors

Jörn Becker
BGZ Gesellschaft für Zwischenlagerung (BGZ Company for Interim Storage), Germany

David Brazier
The Environment Agency, United Kingdom

Anne Eckhardt
risicare GmbH, Switzerland

Jo-Ann Facella
Nuclear Waste Management Organization (NWMO), Canada

Armin Grunwald
Nationales Begleitgremium (NBG, National Citizens' Oversight Committee), Germany

Mathias Häring
BGZ Gesellschaft für Zwischenlagerung (BGZ Company for Interim Storage), Germany

Karena Kalmbach
Eindhoven University of Technology, The Netherlands

Wolfgang Kröger
Eidgenössische Technische Hochschule (ETH) Zürich, Switzerland

Pascale Jana Künzi
Swiss Federal Office of Energy (SFOE), Switzerland

Michael I Ojovan
Imperial College London, United Kingdom

Konrad Ott
Christian-Albrechts-Universität zu Kiel, Germany

Klaus-Jürgen Röhlig
Technische Universität Clausthal, Germany

Erik Setzman
Svensk Kärnbränslehantering AB (SKB, Swedish Nuclear Fuel and Waste Management Co), Sweden

Andrej Stankovski
Eidgenössische Technische Hochschule (ETH) Zürich, Switzerland

Maik Stuke
BGZ Gesellschaft für Zwischenlagerung (BGZ Company for Interim Storage), Germany

Clemens Walther
Institut für Radioökologie und Strahlenschutz, Leibniz Universität Hannover, Germany

Hans Wanner
Former WENRA Chair

Part I

Scientific and technical background

IOP Publishing

Nuclear Waste
Management, disposal and governance
Klaus-Jürgen Röhlig

Chapter 1

Radioactivity

Clemens Walther

Radioactivity is a natural phenomenon. Man and all living species on earth are, and always were, exposed to ionizing radiation. While the natural level of radiation exposure varies within roughly one order of magnitude for different regions on earth, the mean global values might serve as a good standard of what has to be considered 'normal'. Regardless of whether this level of radiation exposure should be called harmless, since some diseases might originate from exposure to naturally occurring radiation, this exposure is unavoidable. Hence, the principles of radiation protection, such as dose limits or constraints, are based on and compared to this natural exposure.

Natural sources of ionizing radiation are responsible for the majority of radiation exposure of the world's population (Cinelli *et al* 2019). Anthropogenic activities dominate only in highly contaminated areas, e.g. due to accidental release from nuclear facilities, nuclear weapon detonations or mining activities (Eisenbud 1997). Discharges from nuclear installations in regular operation are generally rather low nowadays and cause maximum doses to members of the public that are typically two orders of magnitude below the natural background. Medical uses of radioactivity and ionizing radiation in industrial countries cause mean doses comparable to those of natural radiation; however, individual doses differ considerably.

This chapter starts with a short overview of the sources of natural radionuclides and the resulting dose to man. The following section describes the general radiation protection principles as recommended by the International Commission on Radiological Protection (ICRP), including widely accepted dose limits and constraints for members of the public and workers. The application of these general principles to the special case of radioactive waste disposal and clearance concludes the chapter. Important terms and definitions of radiation protection terms are written in *italics*.

doi:10.1088/978-0-7503-3095-4ch1

1.1 Primordial, radiogenic, and cosmogenic radioactivity and ionizing radiation

Natural radioactivity and radiation have two primary causes: on the one hand, high-energy cosmic radiation interacting with the upper atmosphere and, on the other hand, long-lived (primordial) radionuclides in the Earth's crust and their decay products (terrestrial radiation). The most important kinds (sometimes called qualities) of radiation are: *alpha radiation* (helium nuclei emitted by heavy nuclei), *beta radiation* (electrons or positrons emitted by nuclei), *gamma radiation* (highly energetic photons emitted by excited nuclei), *neutrons* (emitted, for instance, by the fission of nuclei), and also *x-rays*, photons emitted from the electronic shell of an atom, or produced by the acceleration (or deceleration) of charged particles. Alpha and beta radiation are stopped by rather thin layers of matter, whereas neutrons and gamma radiation are more penetrating. For more details see e.g. Kratz and Lieser (2013) and also box 1.1.

Cosmic radiation consists, for the most part, of energetic protons and helium nuclei, as well as gamma radiation. They interact with the components of the Earth's upper atmosphere and produce secondary radiation, mainly photons and muons; at higher altitudes, they also produce neutrons, as well as cosmogenic radionuclides, e.g. ^{3}H, ^{7}Be and ^{10}Be, ^{14}C and ^{129}I.

The most important primordial radionuclides are ^{40}K, ^{238}U, ^{235}U, and ^{232}Th and their decay products (radiogenic nuclides, e.g., ^{226}Ra, ^{210}Po, ^{210}Pb). ^{40}K plays by far the most important role, since potassium is an essential element for humans, and

Box 1.1. ***Nuclides, radionuclides, isotopes, symbols***

A *nuclide* consists of atoms with unambiguously defined numbers of protons and neutrons; it has a defined energy level.

If a nuclide is unstable, i.e. its nuclei change over time, and is thus radioactive, i.e. the nuclei emit radiation during these changes, it is called a *radionuclide*.

Nuclides belonging to the same element (and therefore consisting of atoms with the same number of protons) but possessing different numbers of neutrons per nucleus (and therefore different atomic masses) are called *isotopes*.

As a consequence, isotopes behave similarly from a chemical perspective, but their (nuclear) physics varies.

In order to distinguish isotopes, not only the name of the element but also the mass of the nucleus has to be named, e.g. ^{14}C or C-14. 'C' denotes that the element in question is carbon (the number of protons is six); '14' denotes the mass of the nucleus (which thus contains eight neutrons).

Sometimes, symbols such as $^{14}_{6}C$ or even $^{14}_{6}C_{8}$ are used, but they obviously carry no additional information.

If nuclei contain equal numbers of protons and neutrons but contain different energy levels (so-called 'nuclear isomers'), this can be expressed by using an 'm' for the higher energy level ('excited state'), e.g. by ^{242m}Am or Am-242m (as opposed to ^{242}Am).

thus radioactive ^{40}K is always present in the body at a proportion of about 0.012% of the potassium content. The equally primordial nuclides ^{238}U, ^{235}U, and ^{232}Th, unlike ^{40}K, do not decay into stable daughters. They form long decay series of radioactive nuclei, the so-called radiogenic nuclides. Several of them are alpha emitters with correspondingly high relative biological effectiveness (high local damage to living tissue). Some of these nuclides decay within seconds, while others have half-lives in the range of several years. The decay series end at the stable lead isotopes ^{206}Pb, ^{207}Pb, and ^{208}Pb.

The above radionuclides are omnipresent in soils and rocks, but with considerable variability in terms of concentration and also mobility (Atwood 2010). On a global average, dried topsoil contains 370 Bq kg^{-1} ^{40}K, and 35 Bq kg^{-1} each of ^{232}Th and ^{238}U (UNSCEAR 2000). The latter corresponds to the global average concentration of about 3 g of uranium per ton of soil. Some rocks, such as granite, contain many times these concentrations, and chemical leaching, separation, and transport processes can locally enrich certain elements of the decay series. For example, ^{226}Ra is part of the ^{238}U decay series, and much more soluble and therefore more mobile than its parent, ^{230}Th. The daughter of ^{226}Ra is the gaseous ^{222}Rn.

A complete overview of the naturally occurring radioisotopes and the role of radioactive equilibria is beyond the scope of this chapter. For this, see, e.g., Kratz and Lieser (2013) and Magill *et al* (2018).

1.2 A very brief introduction to dosimetry

In order to quantify the risk or potential damage to humans due to exposure to ionizing radiation or following the ingestion of radioactive substances, a well-established dosimetry system is available. The challenge for radiation protection can be demonstrated by the following exemplary requirement. The risks originating from exposure situations as different as *external exposure* to penetrating gamma radiation on the one hand (e.g. ^{40}K or ^{137}Cs) and the *ingestion* of short-range alpha emitters (e.g. ^{210}Po or ^{239}Pu) or the *inhalation* of radioactive radon gas on the other hand need to be compared, weighed, and added. It is neither possible nor necessary to describe all the details of the dosimetric system here. Nevertheless, a short overview of the most important quantities is given in the following. The definitions are taken from ICRPAEDIA (2021). The ICRP defines a radiation dose as a measure of the amount of exposure to radiation. There are three kinds of dose in radiological protection. The absorbed dose is a measurable, physical quantity, while the equivalent dose and the effective dose are specifically conceived for radiological protection purposes.

The *absorbed dose, D,* is the amount of energy deposited in a mass by radiation. D is measured using the unit Gray (Gy = J kg^{-1}). The mass can be, e.g., water, rock, air, human tissue, etc.

The *equivalent dose,* H_T, is calculated for individual organs T. It is based on the absorbed dose to an organ D_T, adjusted to account for the effectiveness of the type of radiation. Radiation weighting factors are needed because different types of radiation (such as alpha, beta, gamma, and neutrons) can have different effects, even if the absorbed dose is the same. For gamma (photon) and beta (electron) radiation, the radiation weighting factor is $w_R = 1$. Therefore, for example, an absorbed dose of 1 mGy in an organ equals an equivalent dose of 1 mSv to that organ. For alpha radiation, $w_R = 20$; for neutrons of unknown energy, $w_R = 10$.

$H_T = \Sigma(w_R D_T)$ is measured in units of Sv = J kg^{-1} (Since w_R is dimensionless, Sv has the same representation in SI units as Gy). More frequently, fractions such as millisieverts (mSv) or microsieverts (μSv) are used. The organ should always be specified (for example '25 mSv to the skin').

The *effective dose, E,* in particular is a central feature of radiological protection and answers the challenge mentioned at the beginning of this section. It sums up any number of different exposures into a single number that reflects the overall risk. The effective dose is calculated for the whole body.

It is the sum of equivalent doses to all organs, each adjusted to account for the sensitivity of the organ to radiation.

$$E = \sum(w_T H_T).$$

The tissue weighting factors w_T are needed because radiation sensitivity is different for each organ, even if the equivalent dose is the same. For instance, the lung ($w_T = 0.12$) is much more sensitive than the skin ($w_T = 0.01$).

The effective dose is also expressed in sieverts (Sv). This is the most frequently used dose in radiological protection. Unless a specific organ is mentioned, a 'dose' in Sv or mSv is the effective dose.

Care has to be taken when calculating the risk from the effective dose. It is very important to keep in mind that the effective dose is not meant to calculate an exact risk for an individual person. The effective dose is merely suited for the assessment of exposure situations and used in practical radiation protection. In very simple words, the effective dose is 'calibrated' versus risk via epidemiological studies (Ozasa *et al* 2019). From the follow-up studies of the atomic bomb survivors in Hiroshima and Nagasaki, the effective dose to ca. 100 000 individuals was related to the probability of suffering from radiation damage. In most cases, this takes the form of solid cancer or leukaemia. From these data, the risk for members of the public, averaged over sexes and over all age groups, is 5.5% per Sv (ICRP103 2007). This kind of radiation damage is called a *stochastic effect.* Even very small doses may lead to an (also small) enhanced risk. Radiation protection aims to minimize this risk (see section 1.4.2).

In addition, there are *deterministic effects* (also called tissue effects) such as acute radiation syndrome, skin burns, or damage to the gastrointestinal system. These effects only occur above a certain threshold of the order of 250–500 mGy (absorbed dose). Proper radiation protection can prevent the occurrence of deterministic effects by keeping doses significantly below this threshold.

1.3 Dose due to natural radioactivity

External exposure to secondary cosmic rays leads to a global average annual effective dose at sea level of 280 μSv due to the direct ionizing component, plus

48 μSv due to neutrons. The dose increases with increasing altitude and latitude. In the mountains, at an altitude of 4000 m, the former's contribution is about four times higher, and the neutron dose can even be 32 times higher (UNSCEAR 2010). The cosmogenic radionuclides mentioned in section 1.1 only play a minor role. They enter the human body through the food chain and result in a mean annual effective dose of 12 μSv.

Potassium is an essential and at the same time omnipresent element. Therefore, it is not surprising that the primordial radioactive isotope ^{40}K contributes a large fraction to the ingested dose. Foodstuffs often contain quite high potassium contents and thus high active concentrations of ^{40}K. However, the potassium concentration is kept constant by healthy humans within quite narrow limits (in all tissues, it is just under 0.2%). This results in a likewise quite constant specific activity in the human body of about 60 Bq kg^{-1} corresponding to an annual effective dose of about 180 μSv, using the conversion coefficient of 3 μSv per Bq kg^{-1} (NCRP 1987).

Of the remaining radionuclides, ^{210}Pb, ^{210}Po, ^{226}Ra, ^{228}Ra, ^{238}U, ^{230}Th, ^{228}Th, ^{232}Th, and ^{235}U are the radionuclides with the highest ingestion rates from the total diet on a global average. The annual ingested activity for many countries follows log-normal distributions with a width of more than one order of magnitude. Medians for the above radionuclides range from about 0.2 Bq for ^{235}U and ^{232}Th to 30 Bq and more than 50 Bq for ^{210}Pb and ^{210}Po, respectively (UNSCEAR 2000).

Drinking water is often the main source of uranium and radium. The marine animals fish, mollusks, and crustaceans are major sources of ^{210}Po, with representative concentrations of 2.4 Bq kg^{-1}, 15 Bq kg^{-1}, and 6 Bq kg^{-1}. The consumption of marine animals is very country (or region) specific. A consumption of about 13 kg per year results in a mean intake of about 30 Bq of ^{210}Po (UNSCEAR 2000).

After incorporation into the human body, the radionuclides partly accumulate quite specifically in certain organs or tissue types: thus, Ra (as a homologue of Ca) and Pb are preferentially found in the bones; U is additionally found in the kidneys. The transfer of Po occurs preferentially in the liver and kidney and that of Th occurs in bones (surface and medulla), the liver, and the gonads. K is located in muscle tissue and is therefore distributed quite homogeneously in the body. From these data, effective doses are obtained using biokinetic models.

The ingestion of primordial radionuclides results in an annual effective dose of 290 μSv and external exposure caused by their gamma radiation results in an annual effective dose of 480 μSv (with a variability of 300–1000 μSv.) The gaseous radiogenic nuclides ^{222}Rn and ^{220}Rn contribute by far the largest part to the inhalation dose, namely, an effective dose of 1250 μSv per year (with a variability of 250–5000 μSv).

For all these mean values, the considerable global variability has to be kept in mind: in different areas of the world, contributions can vary by more than one order of magnitude, adding up to annual effective doses ranging from just under 1 mSv in the South Pacific islands to more than 50 mSv in parts of India, Brazil, and Iran.

A very recommendable overview of the topics summarized in sections 1.1–1.3 is available in the booklet 'Radiation—Effects and Sources' produced by the United Nations (UNEP 2016).

1.4 Radiation protection

In contrast to the unavoidable exposure due to natural radioactivity, anthropogenic use of ionizing radiation and radioactive substances is strictly regulated. The ICRP distinguishes planned exposure situations, existing exposure situations, and emergency situations (see section 1.4.1). International standards require justification, optimisation, and the application of dose limits. Depending on the kind of exposure situation, different regulatory dose limits or constraints apply (see section 1.4.2). These limits differ for workers and members of the public. This chapter refers to the international standards suggested by the ICRP. In addition, national and transnational regulations exist. For the member states of the European Union, a framework based on the ICRP recommendations was given in the Basic Safety Standards (BSS) (BSS 2014). Most member states subsequently adapted their national regulations to comply with these.

1.4.1 Exposure situations

Sources of ionizing radiation might be natural or artificial (manmade) in origin. The human body does not distinguish between these two; the potential damage only depends on the kind of exposure with respect to radiation (section 1.1) and dose (section 1.2). However, regulation is very different. Exposure to natural radiation, if not enhanced by human activity, is not regulated. Exposure to radiation originating from or enhanced by human activity is regulated. In this case, three so-called exposure situations (adapted from ICRP103 2007 and ICRP138 2018) are introduced:

The *planned exposure situation* results from the deliberate introduction and operation of radiation sources, used for their radioactive properties. For this type of situation, the use of the source is understood, and as such the exposures can be anticipated and controlled from the beginning. One example is a nuclear installation used for the production of medical isotopes. During the planning phase, one can assess planned discharges resulting in ingested or inhaled doses to the public due to traces of radionuclides in water or the air and direct radiation (e.g. penetrating gamma radiation), which causes an *external dose* and doses to workers during regular operation. This needs to be justified (see section 1.4.2). The license for building the plant will only be granted if dose limits are complied with; during operation, this will be continuously monitored by the authorities.

An *existing exposure situation* results from a source that already exists, with no intention to use the source for its radioactive properties, before a decision to control the resulting exposure is taken. Decisions on the need to control the exposure may be necessary but not urgent. One example can be taken from the legacies of former uranium mining. The originator might no longer exist. The mine and the radioactive substances dispersed on the ground are no longer used. However, people are living close to contaminated tailing piles and might be exposed to enhanced levels of gamma radiation and radon concentration in air. Remediation might be necessary, but there is no longer a way to prevent the contamination by advance planning as in the planned exposure situation. Furthermore, there is, in general, no benefit to society to be weighed against the exposure, as needs to be done in the first case (see section 1.4.2).

An *emergency exposure situation* results from the loss of control of a source, or from intentional misuse of a source, which requires urgent and timely actions in order to avoid or mitigate exposure. The most dramatic example is a severe accident in a nuclear power plant during operation. In this case, regaining control of the situation and preventing death or injury to workers and members of the public has absolute priority. Immediate action is necessary to limit the dose to workers and the population (so-called averted dose). A different dose regulation from that of planned exposure situations applies, still based on ICRP60 (1991) with adaptions published in ICRP109 (2009). For members of the public, so-called reference levels for the projected dose are defined instead of dose limits; for emergency workers, doses exceeding those for planned exposure situations might be tolerable. Details are given in ICRP109 (2009).

1.4.2 Principles of radiation protection and dose limits

The ICRP defines the three basic principles of radiation protection in its publications ICRP103 (2007) and ICRP123 (2013):

The Principle of Justification: 'Any decision that alters the exposure situation should do more good than harm.' This principle is subject to societal judgement and may change over time. Whereas there is little doubt that an x-ray investigation prior to surgery on a broken bone is a justified exposure, this is not so clear for the use of nuclear power for electricity generation. Some countries rate the benefits for society as high enough to outweigh the risks, while others do not. Moreover, countries phasing out (or in) nuclear power also illustrate that the outcome of the (political) evaluation of justification may change over time.

The Principle of Optimisation of Protection: 'The likelihood of incurring exposure, the number of people exposed, and the magnitude of their individual doses should all be kept as low as reasonably achievable, taking into account economic and societal factors.' This principle, often referred to as the ALARA principle (as low as reasonably achievable), requires the minimization of the radiation-induced risk to persons. This is also true for doses below the regulatory limits. However, it does not imply the reduction of risk (or dose) to ever lower values no matter the cost. First, zero risk can never be achieved, and second, a further reduction of already small risks (or doses) should only be required at 'reasonable' cost or effort. The ICRP (ICRP104 2007) defined a so-called 'trivial dose' of a few tens of μSv per year (effective dose). This might act as an orientation for what is to be considered a small dose.

The Principle of Application of Dose Limits: 'The total dose to any individual from regulated sources in planned exposure situations other than medical exposure of patients should not exceed the appropriate limits specified by the Commission.'

It is important to note that in this case the ICRP refers to planned exposure situations only (see section 1.4.1). Later on in this chapter, we will discuss whether the disposal of nuclear waste in general and a deep geologic repository for radioactive waste in particular are planned exposure situations. Hence, it is appropriate to define dose limits for members of the public and for workers. The internationally most accepted dose limits are as follows: for workers, the annual effective dose should not exceed 20 mSv (if it is exceeded in one year, 100 mSv must not be exceeded in 5 years). The lifetime dose limit is 400 mSv (not including natural background radioactivity). In emergency situations, emergency workers may receive doses generally not exceeding 100 mSv, but in exceptional situations, up to 500 mSv (BSS 2014).

For members of the public, the additional annual effective dose due to anthropogenic actions (from planned exposure situations) may not exceed 1 mSv. The latter limit is to be understood to be the sum of all sources from planned exposure situations including all exposure pathways (discharges into the air, into the water, and external exposure due to penetrating radiation, respectively). For a conservative dose estimation, the exposure of a so-called *representative person* is considered. This is 'an individual receiving a dose that is representative of the more highly exposed individuals in the population' (ICRP101 2006). However, the regulator or authority may set lower limits for a single source (for instance, for the dose due to releases from a nuclear installation).

1.5 Radiation protection aspects of nuclear waste disposal

In several volumes published on the radiation protection aspects of nuclear waste disposal (ICRP77 1997, ICRP81 1998, ICRP122 2013), the ICRP considers nuclear waste disposal to be a part of the practice that generates the waste. Nuclear power generation is a planned exposure situation (see section 1.4.1) and consequently so is handling its waste. Hence, regulations and guidelines for planned exposure situations apply entirely to the potential exposure to humans and the environment associated with the operation and the expected evolution of the geological disposal of long-lived solid radioactive waste (ICRP122 2013). This reasoning remains valid for most waste that needs to be disposed of today, even if it is caused by other actions than nuclear power generation. Building and operating a deep geological repository as well as installations to handle or condition the waste are clearly also planned exposure situations.

The reasoning above remains valid for all categories of radioactive waste. However, for low-level waste, so-called *clearance* is an alternative concept in some countries.

The IAEA defines *clearance* as the removal of regulatory control by the regulatory body from radioactive material or radioactive objects within notified or authorised facilities and activities (IAEA 2019). General exemption and clearance criteria can be found in paragraph 3 of Annex VIII of the BSS (2014). The general criteria for the clearance of materials from authorised practices are that (i) the radiological risks to individuals caused by the practice are sufficiently low as to be of no regulatory concern, (ii) the type of practice has been determined to be justified, and (iii) the practice is inherently safe. The BSS (2014) define exemption levels, i.e. isotope specific maximum activity values or activity concentration values. When handling substances with radioactivity below these levels, practices may be exempted from notification to or regulation by the competent authority. The BSS also recommend using these criteria for clearance. In close correspondence with the BSS, the IAEA derives generic clearance levels from the requirement that 'annual doses to individual members of the public arising from any single cleared practice will not exceed 10 μSv' (TECDOC-1000 1998). Some countries, for example, Germany, do not exactly follow this approach but rather require proof that an annual effective dose 'in the range of 10μSv' is not exceeded for members of the public (STRLSCHV18 2018). As a consequence, the activity concentration levels for clearance may differ from those for exemption.

Considering waste management and disposal operations as planned exposure situations, the justification principle applies (see section 1.4.2). This justification should be reviewed over the lifetime of that practice whenever new and important information becomes available. The justification must include all actions including (multiple) transport of the waste, buffer storage, reconditioning, operation of the repository and, if applicable, any options necessary for recovery (or retrieval). The optimisation of protection is one of the central elements of the stepwise design, construction, operation, and closure of a repository.

1.5.1 Dose limits

Employees in a repository are occupationally exposed workers in the sense of ICRP103 (2007). As in any nuclear installation, their dose needs to be systematically monitored based on individual measurements performed by a dosimetry service (BSS 2014). For all phases before the repository is finally closed (the operational phase), their effective dose must remain below the limit of 20 mSv per year or 100 mSv in 5 years.

While the dose to occupationally exposed workers only needs to be considered until the sealing of the repository, the potential dose to members of the public needs to be optimised far into the future. For planned exposure situations, the radiological criterion for the design of a waste disposal facility recommended by ICRP is an annual dose constraint for the population of 0.3 mSv year (ICRP122 2013), calculated over all exposure pathways as described in (section 1.4.2).

However, the very long timescales that need to be considered in the case of a repository for long-lived radioactive waste complicate the direct transfer and application of the radiation protection regulations and the principles of 'regular' planned exposure situations. Furthermore, a repository may evolve along different scenarios (NEA 2001, 2015). In a probabilistic approach, each of these scenarios is assigned a certain probability and a dose to the representative person. ICRP suggests the use of a risk constraint for the population of 1×10^{-5} per year when applying an aggregated approach that combines the probability of the exposure scenario and the associated dose. Formally, an effective dose constraint might be calculated using the nominal risk coefficient of 5.5% Sv^{-1} for cancer (ICRP103 2007). However, one needs to keep in mind that the validity of the underlying model (linear, no threshold) and hence the applicability of this risk coefficient for doses below 10 mSv is not yet proven. Furthermore, the underlying data are sex and age averaged and were obtained from a very short high-dose-rate exposure situation (atomic bomb survivors). The exposure of members of the public due to discharge or release from a repository for radioactive waste will, in contrast, be at a (very) low dose rate, but involve long-term exposure.

National regulations might differ and evolve over time (NEA 2007). The German Commission on Radiological Protection (SSK) and the Reactor Safety Commission (RSK) recommended the use of the following dose criteria: an annual effective dose of 0.1 mSv for probable evolution scenarios and 1 mSv for less likely scenarios (SSK/RSK 2008). With reference to the *trivial dose* (ICRP104 2007), the German Federal Ministry for the Environment, Nature Conservation and Nuclear Safety (BMU) changed the terminology and tightened the requirement by aiming at an annual effective dose to members of the public 'in the range of 10 μSv' for expected evolution scenarios and a maximum of 100 μSv for deviating scenarios (EndlSiAnfV 2020).

1.5.2 The long-term assessment of a geological disposal facility

Due to the issues mentioned in section 1.5.1, the ICRP recommends understanding optimization in the broadest sense as an iterative, systematic, process (ICRP122 2013). Three main time frames have to be considered: (i) the time of direct oversight, when the disposal facility is being operated and is under active supervision; (ii) the time of indirect oversight, when the disposal facility is partly or fully sealed, in which indirect regulatory, administrative, or societal oversight might continue; and (iii) the time of no oversight, when the memory of the disposal facility has been lost. The application of the protection system is clearly influenced by the level of oversight or 'watchful care' of the disposal facility. Time frames (i) and (ii) can be handled rather similarly to the operation of nuclear installations today. This is no longer the case when oversight ceases to exist in the post-closure period. Nevertheless, the disposal system is still a functioning facility and potential exposures should be considered as planned.

Although regulatory control of a geological disposal facility is not envisaged to continue indefinitely, the disposal of long-lived hazardous waste in a geological disposal facility is a totally different concept to the exemption or clearance of waste from regulatory control, and hence the dose criteria for clearance do not apply (ICRP122 2013).

For the protection of future generations, the ICRP recommends continuous reliance on the basic principle that: 'individuals and populations in the future should be afforded at least the same level of protection as the current generation' (ICRP81 1998). An assessment of the effective dose, as described in (section 1.4.2), is not possible, due to the high uncertainties with respect to the evolution of the repository, future release rates from the installation, radioecological biosphere modelling, and the living habits of future populations. Rather, the use of the public exposure dose limits must be replaced by a process of constrained optimisation in the control of radioactive waste disposal (ICRP77 1997). In the very long term, dose and risk criteria should be used for the comparison of disposal options rather than as a means of assessing health risks.

The risk or dose constraints do not apply for severe natural disruptive events not taken into account in the design-basis evolution and inadvertent human intrusion.

If these events take place while there is still (direct or indirect) oversight of the disposal facility, the ensuing exposure situation (emergency or existing) should be considered by the competent authority, and the relevant protection measures should be implemented (ICRP122 2013). For emergency exposure situations, the ICRP recommends a reference level in the range of 20–100 mSv, and the development of protection strategies to reduce exposures to as low as reasonably achievable below the reference level, taking economic and societal factors into account (ICRP109 2009, Kai 2016). According to ICRP103 (2007), long-lasting exposures resulting from natural disruptive events (with or without an emergency phase) should be referred to as 'existing exposure situations', and the recommended reference level for optimising protection strategies ranges between 1 and 20 mSv $year^{-1}$. In agreement with (Lochard 2012), a reference level should be selected in the lower part of the band (e.g. in the range of a few mSv $year^{-1}$). These regulations are tailored for, and adapted from, severe accidents at operational nuclear power plants. It is a very unlikely scenario that a deep geologic nuclear waste repository in the post-closure phase could give rise to an emergency situation. Consequently, the same is true for a subsequent existing situation of even approximate dimensions, as in the case of, e.g., the Fukushima accident.

Bibliography

Atwood D A 2010 *Radionuclides in the Environment* (New York: Wiley-Blackwell)

BSS 2014 Council of the European Union European Council Directive 2013/59/Euratom on basic safety standards for protection against the dangers arising from exposure to ionising radiation and repealing Directives 89/618/Euratom, 90/641/Euratom, 96/29/Euratom, 97/43/Euratom and 2003/122/Euratom *Off. J. EU* **L13** 1–73

Cinelli G, Tollefsen T, Bossew P, Gruber V, Bogucarskis K, De Felice L and De Cort M 2019 Digital version of the European Atlas of natural radiation *J. Environ. Radioact.* **196** 240–52

Eisenbud M 1997 *Environmental Radioactivity from Natural, Industrial and Military Sources* 4th edn (New York: Academic)

EndlSiAnfV (Endlagersicherheitsanforderungsverordnung) 2020 *Verordnung über Sicherheitsanforderungen an die Endlagerung hochradioaktiver Abfälle (in German, Ordinance on Safety Requirements for the Disposal of High-Level Radioactive Waste)* (non-binding translation by the German Federal Ministry for Environment, Nature Conservation and Nuclear Safety BMU) http://www.gesetze-im-internet.de/endlsianfv/ [accessed 10 April 2021]

IAEA 2019 *Safety Glossary: 2018 Edition* (Vienna: IAEA)

ICRP60 1991 ICRP Publication 60: 1990 Recommendations of the International Commission on radiological protection *Ann. ICRP* **21**

ICRP77 1997 ICRP Publication 77: Radiological protection policy for the disposal of radioactive waste *Ann. ICRP* **27**

ICRP81 1998 ICRP Publication 81: Protection recommendations as applied to the disposal of long-lived solid radioactive waste *Ann. ICRP* **28**

ICRP101 2006 ICRP Publication 101a: Assessing dose of the representative person for the purpose of radiation protection of the public *Ann. ICRP* **36** 5–62

ICRP103 2007 ICRP Publication 103: The 2007 recommendations of the International Commission on radiological protection *Ann. ICRP* **37** 1–332

ICRP104 2007 ICRP Publication 104: Scope of radiological protection control measures *Ann. ICRP* **37**

ICRP109 2009 ICRP Publication 109: Application of the commission's recommendations for the protection of people in emergency exposure situations *Ann. ICRP* **39** 1–110

ICRP122 2013 ICRP Publication 122: Radiological protection in geological disposal of long-lived solid radioactive waste *Ann. ICRP* **42** 1–57

ICRP123 2013 ICRP Publication 123: Assessment of radiation exposure of astronauts in space *Ann. ICRP* **42** 1–339

ICRP138 2018 ICRP Publication 138: Ethical foundations of the system of radiological protection *Ann. ICRP* **47** 1–65

ICRPAEDIA 2021 http://icrpaedia.org

Kai M 2016 Update of ICRP Publications 109 and 111 *Health Phys.* **110** 213–6

Kratz J V and Lieser K H 2013 *Nuclear and Radiochemistry: Fundamentals and Applications* 3rd rev edn (New York: Wiley-VCH)

Lochard J 2012 Protection of people living in long-term contaminated areas after a nuclear accident: the guidance of ICRP Publication 111 *J. Radiol. Prot.* **32** N95–9

Magill J, Dreher R and Sóti Z 2019 Karlsruhe Nuclide Chart—New 10th edition 2018 *EPJ Nuclear Sci. Technol.* **5** 6

NCRP 1987 National Council on Radiation Protection and Measurements. Exposure of the population in the United States and Canada from natural background radiation *NCRP Report* No 49

NEA 2001 Scenario Development Methods and Practice *OECD NEA An evaluation based on the NEA Workshop on Scenario Development Madrid, May 1999*

NEA 2007 *Regulating the Long-term Safety of Geological Disposal Towards a Common Understanding of the Main Objectives and Bases of Safety Criteria* NEA No 6182

NEA 2015 *Scenario Development Workshop Synopsis—Integration Group for the Safety Case* NEA/RWM/R(2015)3 https://oecd-nea.org/jcms/pl_19690/scenario-development-workshop-synopsis-integration-group-for-the-safety-case

Ozasa K, Cullings H M, Ohishi W, Hida A and Grant E J 2019 Epidemiological studies of atomic bomb radiation at the radiation effects research foundation *Int. J. Radiat. Biol.* **95** 879–91

SSK/RSK 2008 *Gemeinsame Stellungnahme der RSK und der SSK zum GRS-Bericht „Sicherheitsanforderungen an die Endlagerung hochradioaktiver Abfälle in tiefen geologischen Formationen"* Stellungnahme der Reaktor-Sicherheitskommission und der Strahlenschutzkommission

STRLSCHV18 2018 Ordinance on the Protection Against Damage due to Ionising Radiation (Radiation Protection Ordinance) Strahlenschutzverordnung vom 29. November 2018 (BGBl. I Nr 41; 2018 I S. 2034)

TECDOC-1000 I 1998 *Clearance of Materials Resulting from the Use of Radionuclides in Medicine, Industry and Research* (Vienna: IAEA)

UNEP 2016 *Radiation: Effects and Sources*

UNSCEAR 2010 *Sources and Effects of Ionizing Radiation, United Nations Scientific Committee on the Effects of Atomic Radiation* UNSCEAR 2008 Report Vol 1 (Vienna: United Nations)

UNSCEAR 2000 *Sources and Effects of Ionizing Radiation, United Nations Scientific Committee on the Effects of Atomic Radiation* UNSCEAR 2000 Report to the General Assembly with Scientific Annexes Vol 1 (Vienna: United Nations)

Chapter 2

Nuclear fuel from cradle to grave: existing variants and future options for the fuel cycle and resulting waste types

Wolfgang Kröger and Andrej Stankovski

Technology assessment is geared to the three dimensions of sustainability with their associated indicators, and considers the whole life cycle, including all upstream and downstream processes. This kind of standard approach has been applied to nuclear energy and related fuel cycles, and has demonstrated merits such as low greenhouse gas (GHG) emissions and materials flows, including fuel and waste amounts; the drawbacks are the potential for rare but high-consequence accidents and high-level radioactive waste (HLW) which requires long husbandry times. Among nuclear energy's inherent features of specific interest are the incomparably high energy density of uranium and its high assured reserves; many of the next-generation reactor types (GEN-IV reactors, small modular reactors (SMRs)) can use thorium, which is even more abundant and allows for lower amounts of HLW.

The fuel cycle options have been reduced to open cycle, mono, and multi recycle, which can be enhanced by the partitioning and transmutation of minor actinides. All require deep geological disposal, but their requirements differ significantly. In the case of enhanced multi recycle, the time required for the reduction of radiotoxicity to that of natural uranium shrinks to roughly 400 years, compared to 250 000 years for spent fuel in an open cycle.

Effective multi-recycle options need fast reactors. Designs, even those that could be called 'exotic,' which include spent light water reactor (LWR) fuel and an actinide burner, using liquid metals or molten salts as a coolant instead of water, are under development in many countries. Other ways of increasing fuel efficiency and relieving waste management issues, for example striving for higher burnups, have also been introduced. Technology readiness levels vary from high to low, as shown in table 2.1.

doi:10.1088/978-0-7503-3095-4ch2

Table 2.1. Technology maturity of advanced reactor designs (Buongiorno *et al* 2018).

Technology	Maturity
Lead-cooled fast reactors (LFR) with nitride fuel, gas-cooled fast reactors (GFR), and molten salt fast reactors (MSR) that are fast and thermal and use salts other than FLiBe.	Low
Advanced sodium-cooled fast reactors (SFRs), fluoride-salt-cooled high-temperature reactors (FHR), thermal MSRs using FLiBe, LFRs with oxide fuel, very-high-temperature reactors (VHTR) with 900 °C outlets.	Low to moderate
Small conventional SFR and modular high-temperature gas-cooled reactors (HTGR) with 750 °C outlets.	Moderate to high

2.1 The concept of life-cycle analysis and sustainability metrics as applied to nuclear energy

To overcome the limitations of technology assessments that are based on single factors, 'sustainability' has become a standard for the evaluation of technologies; this is also true for the energy sector. There is no single, generally accepted definition of the term, but in Europe at least, a concretization and operationalization via the three dimensions of 'economy,' 'environment,' and 'society' and their associated indicators for the evaluation of energy sources and related technologies has largely prevailed. In each case, a *life-cycle assessment/analysis* (LCA) methodology has to be applied, which means that needs and impacts of all stages of the generation of a commercial product like electricity must be considered. That includes upstream processes from raw material extraction and processing ('cradle') through product (plant) manufacture and use (operation) to downstream processes such as the decommissioning and recycling or final disposal of the materials used ('grave'). A nuclear fuel cycle covers the activities related to fueling nuclear reactors and managing used nuclear fuels. The following discussion from this point onwards refers to Western-designed LWRs, as they are the most common reactor types in the world. The primary fuel cycle starts with the steps of mining and milling uranium, conversion, enrichment, and fuel fabrication, called the 'front end'. After the fuel has been irradiated for three to five years in a reactor to produce electricity, the spent nuclear fuel (SNF) may go through further steps, including interim storage, transportation, and/or reprocessing before SNF is disposed of as high-level waste, known as the 'back end' of the fuel cycle. Uranium in its natural form is composed of 99.3% of the 'fertile' isotope U-238, 0.7% of the 'fissile' isotope U-235, and some traces of U-234. The uranium is processed to produce an 'enriched uranium' product that is about 95% fertile and 5% fissile and a waste stream of 'depleted uranium' that is almost 100% fertile. The enriched uranium is made into ceramic pellets that are placed inside metal 'cladding' tubes. The tubes are arranged in a square lattice that makes up a fuel assembly. These fuel assemblies are placed in the core of the reactor. After most of the fissile atoms are burned and the fissile content drops to a level that is too low to continue, the 'spent' or 'used' nuclear fuel is composed of about 93%

fertile uranium, 1% fissile uranium, and 5% fission products as well as 1% plutonium and 0.1% minor actinides[1]; the latter are created in the reactor when fertile uranium captures neutrons. Most of the fuel content is chemically poisonous and radioactive (NEA 2021). Needless to say, the LCA for the process of electricity production has to take all the 'cradle-to-grave' impacts of the nuclear power plant itself into account, including its construction, operation, and end-of-life activities.

The EU project NEEDS (NEEDS 2016, Hirschberg and Burgherr 2015) represents the current state of the art of such a *sustainability evaluation* framework—based on various methods, notably LCA—with a set of 36 agreed mostly quantifiable indicators. In a nutshell, these are as follows:

- 11 *environmental* indicators that cover energy and mineral resources, climate change, ecosystem impacts from normal operation and severe accidents, and special chemical and medium and high-level radioactive wastes;
- 9 *economic* indicators, including impacts on customers (electricity price), overall economy (employment and autonomy of electricity generation), and utility (financial risk and operation);
- 16 *social* indicators that address the security/reliability of energy provision, political stability and legitimacy, social and individual risks (both expert based (normal operation and accidents) and perceived), terrorist threats, and the quality of the residential environment (landscape and noise).

A broad spectrum of a total of 26 possible advanced power-generation technologies for fossil, nuclear and renewable energy sources was considered, with a view to their medium-term (2030) and long-term operational readiness (2050), taking into account conditions in four countries (France, Germany, Italy, and Switzerland), using today's (2005) technologies as starting points. 'Severe accident risk assessment' is based on worldwide historical data and trend analyses; probabilistic safety analyses (PSA) were used for hypothetical nuclear accidents. The 'social indicators,' which are difficult to quantify, are based on an opinion poll among experts from a wide variety of fields. Some results of special interest are summarized below.

Renewables and nuclear energy produce *GHG emissions* that are one to two orders of magnitude lower than those from fossil technologies without carbon capture and sequestration, see figure 2.1. Radioactive *wastes* are inherently associated with nuclear energy (figure 2.2). The volumes of critical non-radioactive wastes are currently highest for solar PV (45 m^3 stored underground per TWh), but are expected to reduce strongly in the future.

The aggregated *impacts on ecosystems* are the highest for biogas, followed by coal, while the hydropower impacts are lowest[2]. Coal and today's biogas have the highest estimated *health effects*, but there is a high potential for reduction in the case of biogas.

[1] Minor actinides comprise neptunium and elements with atomic numbers beyond that of plutonium, such as americium and curium.

[2] Hydropower is often claimed to have the lowest ecosystem impact. This may be overoptimistic in situations in which the dam totally obliterates all natural life in the flooded valley; the Three Gorges dam in China is an illustration of this.

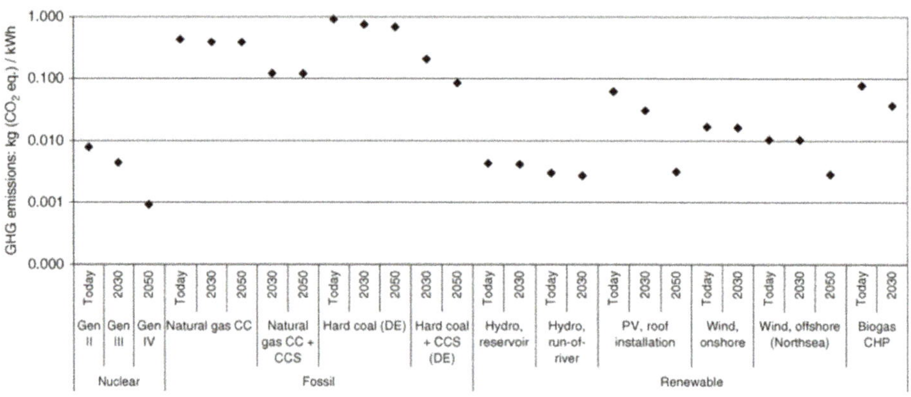

Figure 2.1. LCA-based greenhouse gas (GHG) emissions produced by selected technologies (Hirschberg and Burgherr 2015, John Wiley & Sons). CC stands for combined cycle, CCS for carbon capture and sequestration, PV for photovoltaic, CHP for combined heat and power. Note the logarithmic scale.

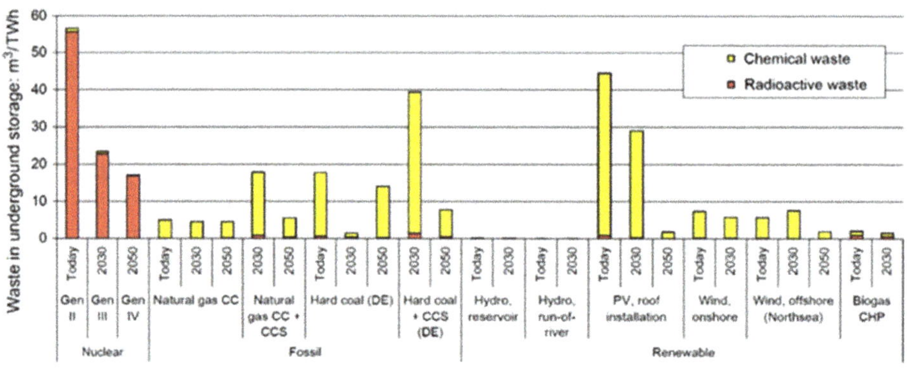

Figure 2.2. LCA-based production of medium and high-level radioactive wastes and special chemical wastes stored in underground repositories (Hirschberg and Burgherr 2015, John Wiley & Sons).

Fossil technologies have the highest expected *risks due to severe accidents* (figure 2.3). The new generations of nuclear reactors exhibit a very substantial reduction of the risk level compared with the current generation. However, the maximum credible consequences of nuclear events, dominated by severe accidents in operated power plants[3], are clearly the highest for nuclear, which leads to a high level of risk aversion.

In addition, large hydropower installations, depending on their location, have the potential to cause very large accidental consequences (dam failures) but without the component of very long-term land contamination, which is a unique feature of nuclear energy. In the case of the most severe nuclear accidents, apart from the

[3] Nuclear fuel cycles have various impacts on public and environmental safety. The greatest (relatively small) risk comes from trucks transporting fuel/fuel feedstock, not due to their content but just due to trucks themselves being on the road. The other potential risk is the long-term risk from long-lived waste that is introduced into the environment (NEA 2021).

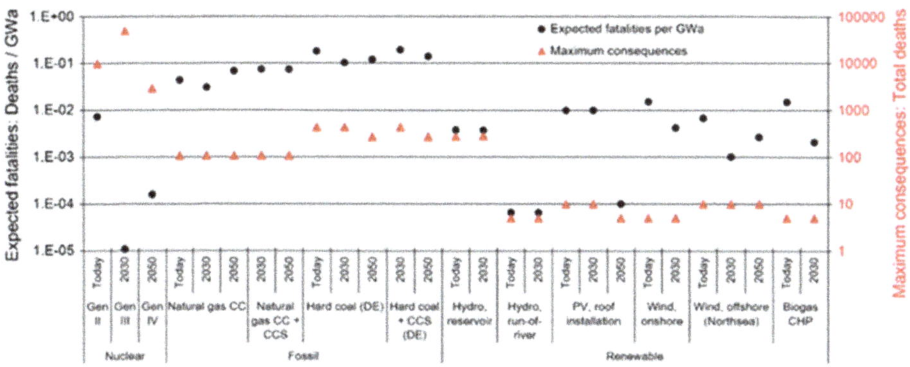

Figure 2.3. LCA-based expected fatality rates due to severe accidents and maximum consequences per accident (Hirschberg and Burgherr 2015, John Wiley & Sons).

evacuation of a large number of people, long-term relocation may be necessary which may cause additional major psychological effects, not considered here.

2.2 Features of specific interest, such as proven fuel reserves and fuel cycle options

One of the most attractive incentives for the use of nuclear fission energy is its incomparably high energy density (80 620 000 MJ kg^{-1} uranium for breeder use, 3.5 million times higher than that of black coal, Sornette *et al* 2018) and related low material flows. From a reliability of supply viewpoint, uranium can be considered a very abundant and easily stored resource. The book *Uranium* 2018*: Resources, Production and Demand* by the NEA and the IAEA (2018) details the proven reserves of natural uranium as 7.9 million tons; the additional inferred and reasonably assured reserves which would be economically viable to extract stand at 7.5 million tons. The yearly consumption of uranium in the 449 commercial reactors operating in 2017 was estimated to be 62 825 tons, suggesting that the current proven reserves can satisfy the current demand for another 125 years, or 254 years if the reasonably assured resources are also taken into consideration. The yearly consumption refers to the total natural uranium demand, including the uranium needed for enrichment, as 80% of all currently operating nuclear reactors are LWRs, which use lightly enriched uranium (3%–5% of U-235). In addition, the currently assured reserves for uranium are higher than what is considered normal for most minerals, indicating that further exploration will likely lead to the discovery of new deposits, albeit at a higher extraction cost.

Even in the most favorable scenarios for nuclear power, which forecast an increase in its worldwide share of electricity production from today's 10.35% to 28% by 2040 (IEA 2019), the increase in demand is less likely to cause a drastic shortage in the uranium supply. History shows that the trends for increased energy efficiency and higher burnup rates have increased the utilization of fuel and stabilized the demand. According to the World Nuclear Association (WNA), in the period from 1980 to 2008 the electricity generated by nuclear power increased by a factor of 3.6, while uranium consumption only increased by 2.5 times (WNA 2020). The discussion so far has been based on the uranium consumed by current LWRs; however, the adoption of new reactor technologies and related fuel cycles, particularly breeder reactors, fuel recycling, and the development of new mining and extraction technologies (e.g. seawater extraction) could theoretically extend the availability of uranium for thousands of years.

Many of the proposed next-generation (GEN-IV, SMR) reactor types can use thorium in the form of the fertile isotope Th-232 as fuel. Thorium is three to five times more abundant than uranium in the Earth's crust and has almost the same energy density (79 420 000 MJ kg^{-1} for breeder use) as uranium. The 2019 IAEA report on thorium (IAEA 2019) estimates the world resources to be approximately six million tons; however, new repositories are bound to be identified when the commercial exploitation of thorium begins. The most important benefits of using thorium compared to uranium are increased proliferation resistance and the smaller amounts of high-level waste (HLW) produced: the thorium fuel cycle yields fewer fission products by up to two orders of magnitude (Moir and Teller 2005). It is important to understand that the thorium fuel cycle has to be started using a combination of uranium and thorium; the used fuel elements are then recycled. The corresponding technology has not yet been developed to an industrial scale (NEA 2021).

The exploitation of nuclear power also has certain drawbacks, both in the physical process and current technologies. The physical process of nuclear fission generates a surplus of neutrons, decay heat (produced by short-lived fission products), radioactive byproducts which are considered to be high-level nuclear waste, etc. This leads to major design challenges, the implementation of safety functions, and the development of comprehensive strategies for the operation and decommissioning of nuclear power plants as well as for the disposal of high- and medium-level nuclear waste (Kröger *et al* 2020).

2.3 Fuel cycle options and their impact on radwaste issues

The relevant issues are (a) the requirement for, and utilization of, natural uranium; (b) heat load and waste volume; (c) the composition and radiotoxicity of waste; and finally (d) safeguarding times, which depend on the fuel and fuel cycles used to different degrees. The main types of fuel cycle include those in which SNF is directly disposed of as waste, recycled once to use the small amount of remaining fissile material, or recycled multiple times in breeder reactors to convert and use the remaining large amount of fertile material. Fissioning the remaining fissile and fertile materials makes more effective use of the mined uranium while removing the source of much of the long-term radioactivity in nuclear waste (NEA 2021).

The *open cycle* is currently considered to be the reference option in most countries that use nuclear energy, mainly due to proliferation and radioactive waste concerns that stem from the current reprocessing techniques. At its heart is a reactor with a thermal neutron spectrum[4], typically cooled by light water (H_2O), known as a LWR. At the back end, the cooled SNF assemblies are safely stored pending their disposal

[4] The majority of fission-generated neutrons are in the high-energy spectrum and are subsequently slowed down to the thermal spectrum using moderation.

in a deep geological repository. The storage may take place at, or away from, the reactor site. With the exception of the packaging and conditioning processes, the SNF will need to comply with transport and disposal requirements and will not receive any particular chemical treatment.

The *mono-recycle* process (sometimes called the partially closed process) is a two-stage process. The first stage is as explained above up to the point at which the cooled SNF is managed. From this stage, it is chemically separated (reprocessed); the uranium and plutonium are recovered for reuse in the second stage, and the minor actinides and fission product HLW is placed into solid, stable waste forms (usually vitrified into glass) and packaged in stainless steel containers. The HLW is stored pending disposal in a deep geological repository. The volume of this waste is smaller than the SNF volume of the open cycle. The fuel hardware (cladding, etc.) is also packaged for disposal as medium-level waste; see figure 2.4.

Extracted plutonium and/or enriched reprocessed uranium (ERU) are mixed with depleted uranium in order to make mixed oxide fuel (MOX) and/or ERU fuel assemblies which are placed back into the reactors. It takes about eight SNF assemblies to recover enough plutonium for one MOX fuel assembly (NEA 2021). The recycling of the plutonium allows the fissile content in the plutonium (around 60%, depending on the burnup history) to be used in fuel. When this MOX fuel is spent, it is placed in an SNF pool for cooling, eventually followed by geological disposal. This fuel cycle only requires about 75% of the natural uranium needed for the open cycle, due to the reuse of the fissile isotopes.

The *multi-recycle* process (sometimes called the closed cycle process) has not been developed at an industrial scale and is the subject of research and development (R&D). It is similar to the mono-recycle fuel cycle, except that the fuel is recycled multiple times; each time, the uranium and plutonium are recovered and the minor actinides and fission products are disposed of.

In general, to obtain enough fissile material for the multi-recycle process, a fast reactor is usually needed for the second stage. This type of reactor uses neutrons with higher energies, which results in more plutonium being made from the fertile uranium, either roughly the same amount of plutonium in the SNF as was in the fresh fuel (for 'break-even'/'converter' reactors) or even more plutonium (for

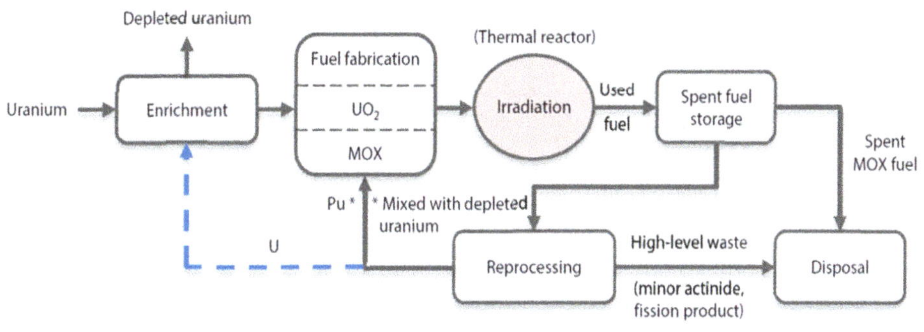

Figure 2.4. The mono-recycle process (Reproduced from NEA 2021, Copyright 2021, with permission from OECD/NEA).

'breeder' reactors). The uranium and plutonium can be recycled continuously, as long as additional fertile uranium is added to each cycle to make up for the loss of fissile material. In addition, the minor actinides and fission products produced during each cycle must be separated and disposed off as waste. The back end of the first and second stages of this fuel cycle is shown in figure 2.5.

Fast reactors can use the plutonium made from fertile uranium or plutonium produced by LWRs using MOX fuel indefinitely, thus extracting most of its energy potential while avoiding the long-term disposal of SNF containing uranium and plutonium. Enhanced recycling concepts include the transmutation of the minor actinides to further reduce the amount of HLW and the time for which it will remain hazardous. These options require additional technologies to separate the SNF into additional streams, to fabricate the fuel, and to transmute isotopes. Fuels containing minor actinides are highly radioactive, which makes the industrial assets under development much more complex; the fuel must be fabricated remotely in special shielded 'hot cells.' In enhanced recycling, strong heat emitters such as cesium and strontium can be separated and stored in dedicated storage containers for several hundred years until the heat from the waste is reduced to a negligible level, significantly reducing the footprint of the final waste repository.

NEA (2021) provides a summary table which differentiates the characteristics of nuclear fuel cycles, from which the following items are depicted here:

- Greater technical challenges are present in the mono-recycle and multi-recycle options, and the enhanced multi-recycle option poses the greatest technical challenge.
- A suitable geology and space are very important characteristics when considering the open cycle. The space required is smaller when the mono- or multi-recycle processes are considered.
- Social acceptance is a common issue for all options, but there are differences that depend on specific situations in certain countries. *(Note: In many countries, reprocessing clearly lacks acceptance or is even forbidden by law.)*

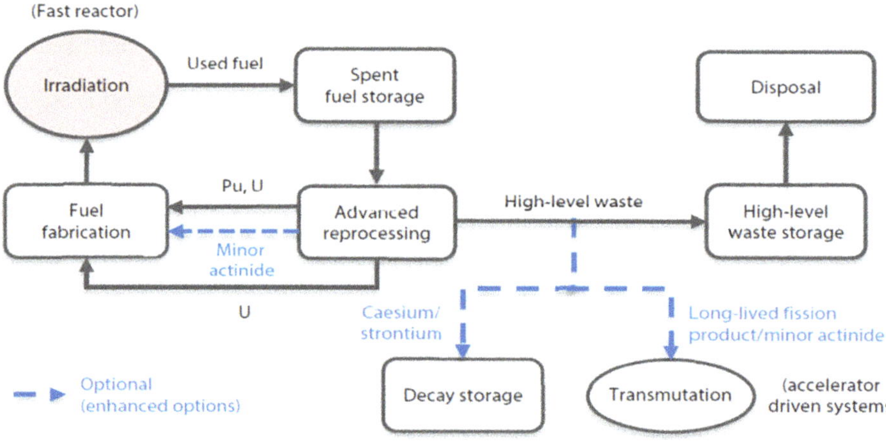

Figure 2.5. Multiple-recycle process (Reproduced from NEA 2021, Copyright 2021, with permission from OECD/NEA).

- Compared to the open-cycle process, the mono-recycle option reduces the natural uranium requirement by up to 25%, while the multi-recycle option dramatically reduces the natural uranium requirement by 99%.
- Reprocessing and the resulting burning of nuclear material in reactors reduces the decay heat, the amount of high-level waste, and the ingestion radiotoxicity.
- Energy independence and security of supply are low for the open-cycle option. If facilities exist nationally, the multi-recycle process is totally self-sufficient. (Only very small amounts of fertile material are needed.)
- The interests of future generations are inherent in all of the differentiating characteristics. Today's decisions determine the opportunities, risks, and challenges for future generations.

2.4 Ways to relax the requirements for radwaste management and required reactor technologies

2.4.1 Higher burnup

One way to relax the requirements for radioactive waste management is to make more efficient use of uranium fuel in existing thermal reactors by *higher burnup*. PWR power reactors operate in a steady state; the composition of the discharged fuel depends on the initial enrichment and the radiation history of the fuel. For a closed fuel cycle, the initial enrichment is typically around 3.5% of U-235 and the average burnup of the assemblies is around 30 to 40 GWd/tHM (gigawatt-days per metric ton of heavy metal). At the end of life (EOL) of the assembly, the fraction of U-235 will be as low as almost 1%. If reprocessing of the used fuel is not foreseen, and thus the isotopic composition of the used fuel does not need to remain in some favorable region for the partitioning process, the burnup can be increased, which generally leads to a more economic use of the fuel. Increasing the burnup will, in turn, reduce the amount of spent nuclear fuel. The change in its isotopic composition is shown in table 2.2.

Today's LWRs use initial enrichments of 4 to 5% of U-235. The limiting factor in reaching higher burnups in PWRs is the in-pile oxidation of the fuel rod cladding.

Table 2.2. Exemplary impact of increasing burnup on the back end of the fuel cycle (Debes 2001).

Impact areas	Case 1	Case 2	Case 3
Burnup (GWd/tHM)	33	45	60
Fission products (kg/TWhe)[5]	140	140	140
Cladding and structural material (kg/TWhe)	1210	890	660
Recyclable uranium (kg/TWhe)	3830	2810	2100
Minor actinides (kg/TWhe)	4.3	4.5	4.7
Recyclable plutonium (kg/TWhe)	37	32	27
Total (1300 MWe LWR, 10 TWhe annually)	52 tHM	38 tHM	29 tHM

[5] TWhe—Terawatt-hour electric.

This cladding is generally made of zirconium-based alloys. During irradiation, the wet side of the cladding becomes oxidized and some of the hydrogen migrates into the cladding material. The outside oxide layer grows to about two-thirds of its thickness inside the cladding material, while the remaining one-third enlarges the original diameter of the fuel rod. The remaining metallic layer of the cladding has to withstand the mechanical forces acting on the cladding, for example, those due to increasing internal radiation pressure. For the original standard cladding, Zircaloy-4, the burnup limits were discussed in the late 1970s (Garzarolli *et al* 1979). The development of new cladding materials to allow higher burnup has included different fabrication processes (the stressed relieved annealed (SRA) and partly and fully recrystallized annealed (RXA) processes). As well as these developments, numerous variations of zirconium alloys have been introduced to reduce the oxidation. These included, for example, the reduction of the tin (Sn) content and the development of composite materials, e.g. Duplex. The latter consists of a Zircaloy-4 rod with a thinner outer layer of a zirconium alloy with a lower Sn content, but higher Fe and Cr contents. The most common materials in Western PWRs are Optimized ZIRLO (developed by Westinghouse) and M5 (developed by Framatome). For further details of the corrosion of the different cladding materials and their chemical and mechanical behaviors, the reader is referred to (Motta *et al* 2015).

Today, fuel rod burnup rates of more than 50 GWd/tHM are seen as standard. High burnup rates of 65 GWd/tHM averaged over the assemblies are implemented e.g. in Germany and Switzerland. To further increase the safety and burnup capabilities of the fuel rods, some actual developments for near-term implementation include chromium-coated claddings, FeCrAl claddings, and e.g. Cr-doped fuel pellets. These developments give rise to the possibility of increasing the burnup limit to 80 GWd/tHM (NEA 2018).

Another type of thermal reactor appears to be more adept in terms of fuel utilization—the high-temperature gas-cooled reactor (HTGR). Some HTGR designs use helium as a coolant and graphite as a moderator, and can reach burnup rates of 60 GWd/tHM. Coolant temperatures of up to 750 °C allow for high thermal efficiency (up to 40%, compared to the average LWR efficiency of 30%–33%), which further increases the utilization of uranium. The increased burnup compared to LWR stems from the design of the fuel elements, which take the form of ceramic pebbles. These pebbles comprise thousands of robust tri-structural isotropic (TRISO) coated particles of fuel embedded in a graphite matrix. The TRISO coated particles consist of a moderately enriched (8.5%) uranium/thorium (UO_2, ThO_2, or UC + ThC) kernel and a silicon carbide (SiC) layer (for fission product retention) in between inner and outer dense pyrolytic graphite layers, which account for fuel expansion (figure 2.6) (Sornette *et al* 2018). The core can be continuously refueled during operation, in which each individual pebble is extracted by the refueling machine, its burnup rate is measured, and it is returned to the core if the rate is below a certain set point. Otherwise, the pebble is discarded and a fresh one is inserted into the core. This level of control and the aforementioned pebble design allow for high burnup rates of the fuel, without the limitations of the zirconium-alloy cladding used in LWRs. Due to the difficulty of recycling and reprocessing the ceramic fuel

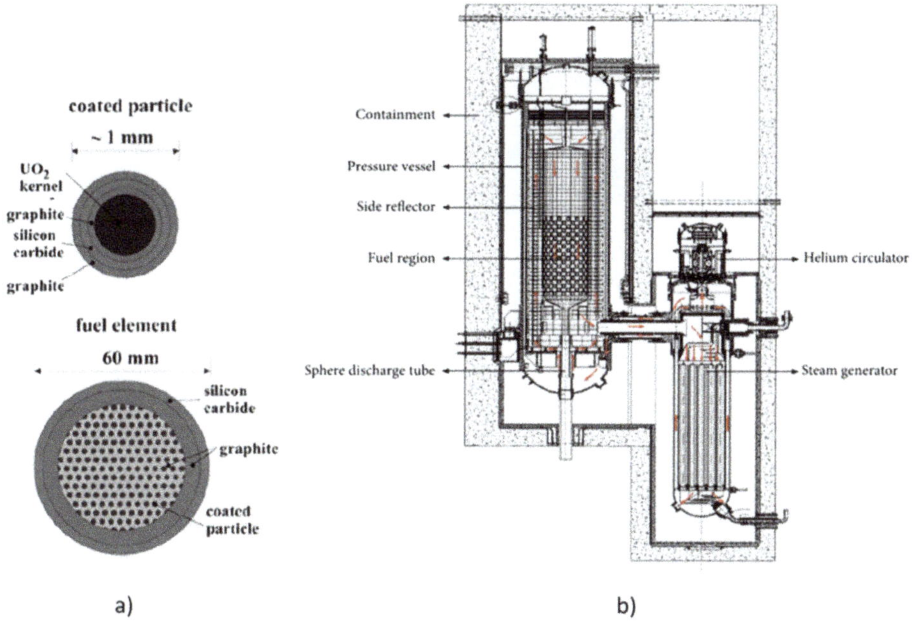

Figure 2.6. Schematic view of HTR—PM: (a) TRISO coated particle and fuel element and (b) layout of a reactor module with 250 MW of thermal power (Reprinted from Hao *et al* 2020, with permission from Hindawi).

elements, the open-cycle process is used for the time being, raising concerns about the comparatively large amount of nuclear waste. Historically, several prototypes employing this technology have been put into operation: AVR (Germany, 1966), THTR-300 (Germany, 1983), and HTR-10 (China, 2003). Currently, the most promising concept of this type is the HTR–PM (pebble bed-modular) reactor in China, which was connected to the electricity grid in December 2021 (Zhang *et al* 2016).

2.4.2 Reduction of husbandry times by the partitioning and transmutation of minor actinides

HLW is highly radioactive, with half-life periods of up to thousands of years. The long-term radiotoxicity mainly originates from plutonium and minor actinides; therefore, the neutralization of these elements using a process known as partitioning and transmutation (P&T) is necessary in order to drastically reduce the husbandry times of nuclear waste. Partitioning refers to the separation of plutonium and minor actinides from spent fuel, while transmutation refers to the conversion of non-fissile actinide isotopes into fissile ones through neutron capture. Finally, the resulting isotopes are fissioned (incinerated, burned) into smaller, less radioactive elements. A study published by the Nuclear Energy Agency (NEA) in 2011 (cf. also NEA 2021) claims that through recovery of minor actinides and their transmutation, the radiotoxicity of spent fuel from LWRs (at a burnup of 51 GWd/tHM) could fall to the level of natural uranium (the baseline value) in 350–400 years, compared to a

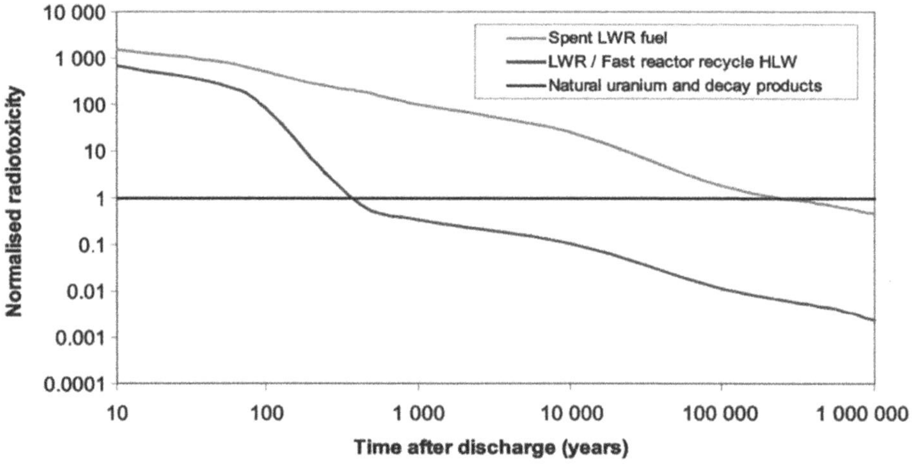

Figure 2.7. Ingestion radiotoxicity for spent LWR fuel (51 GWd/tHM) in both treated (P&T) and untreated cases, normalized to the radiotoxicity of natural uranium and its decay products (Reprinted from NEA 2021, Copyright 2021, with permission from OECD/NEA).

figure of 250 000 years before treatment (figure 2.7). This dramatic reduction will also result in simpler storage conditions, as plutonium and minor actinides are the biggest sources of decay heat in spent fuel. However, the NEA study also states that decreases of estimates of potential radiation exposures to humans in the long-term are not necessarily proportional to radiotoxicity decreases (mainly due to highly mobile fission products which are not transmuted). Besides the additional technologies and expenses involved, a notable downside of P&T is the possible increase in low- and medium-level waste; however, this compromise seems reasonable, as the storage of these types of waste is notably simpler than that of HLW.

Partitioning at a sufficiently high separation rate (currently in the range of 90 to 99.9%) is key and requires the use of complex reprocessing techniques[6] while the most promising transmutation approach is the use of fast breeder reactors, which are capable of burning actinides during normal operation.

These reactors have a very high neutron economy—as a result of utilizing fast neutrons (very low moderation), the fissile atoms (Pu-239 in the U-238 cycle and U-233 in the Th-232 cycle) bred from fertile fuel outnumber the fissile atoms consumed in the process. Most of the fast breeder designs can use fertile fuel mixed with MOX fuel or minor actinides. Such designs currently under development include liquid-metal-cooled fast reactors, molten salt fast reactors, and accelerator-driven systems (Sornette *et al* 2018).

Liquid metal, i.e. sodium- or lead/lead–bismuth-cooled fast reactors have high operating temperatures (>500 °C) and thus high thermal efficiency (~37%), and, most importantly, can burn spent fuel from LWRs. The main shortcomings of these

[6] For example, plutonium uranium recovery by extraction (PUREX), which can separate up to 99.5% of plutonium and minor actinides.

reactor types are the slightly positive void and temperature coefficients, which make them susceptible to reactivity spikes (Sornette *et al* 2018), as well as alleged higher proliferation risks, which necessitate the use of parfait breeding blankets and reprocessing methods that cannot extract plutonium. Sodium fast reactors (SFRs) have lower coolant melting temperatures, but must use an intermediate sodium circuit due to the exothermic reaction between water and the coolant, while lead- and lead–bismuth-cooled reactors (LFRs) do not need an intermediate circuit but have higher coolant melting temperatures leading to freezing concerns. Currently, more than ten liquid-metal reactors are in development, the most promising of which are two small modular reactor concepts due to their use of passive safety systems: PRISM SFR (USA, deployment by 2026, Hitachi *et al* 2018) and BREST-OD-300 lFR (Russia, deployment by 2025, NEI 2019).

Molten salt reactors (MSRs) can achieve very high temperatures (750 °C–900 °C) with higher efficiency, can operate in both the thermal and fast spectrums, and use solid fuel or fuel dissolved in the coolant. Both uranium and thorium-based fuels with added minor actinides can be dissolved in the coolant, making this reactor design ideal for waste burning. The coolant, preferably consisting of lithium salts (due to their high boiling point, >1400 °C), is constantly circulated through the core and a chemical processing plant, in which volatile fission products are separated and the fuel concentration is controlled (Sornette *et al* 2018). This design is resistant to over-heating, as a frozen plug melts above a threshold temperature, dumping the coolant into tanks which are passively cooled, making the reactor safe in station blackout scenarios. The main drawbacks of MSR are the potential criticality spikes and the corrosive properties of the coolant. As MSRs are perfect for waste burning, one such design currently in development is the Seaborg Waste Burner, a small modular reactor that uses thorium and spent fuel from LWRs (Seaborg Technologies 2015).

Accelerator-driven systems (ADSs) are novel concepts designed specifically for the transmutation of minor actinides, and comprise a lead- or lead–bismuth-cooled reactor with an external neutron source, usually a high-intensity proton accelerator (Sornette *et al* 2018). The reactor is subcritical, meaning that it is incapable of self-sustained fission, while the proton accelerator produces neutrons by spallation, sustaining the fission reaction. These systems are resistant to reactivity-induced accidents and do not require the installation of control rods, as the reaction can be stopped by turning off the accelerator. The first large-scale ADS concept, MYRRHA (Multi-purpose Hybrid Research Reactor for High-tech Applications), an actinide burner using 30% enriched MOX fuel, is expected to be commissioned by 2036 by the Belgian Centre for Nuclear Research (figure 2.8) (MYRRHA 2020).

2.4.3 Avoidance of long-lived actinides by the use of thorium (instead of uranium) fuel

Despite the fact that neutrons are moderated to thermal energy in thermal reactors, they can still be captured by U-238, which transmutes it into the unstable U-239, which eventually beta-decays into neptunium Np-239 and finally into fissile Pu-239. There is a one-sixth chance that a captured neutron will transmute a fissile U-235 atom into non-fissile U-236; Pu-239 and Pu-241 have around a one-third chance of

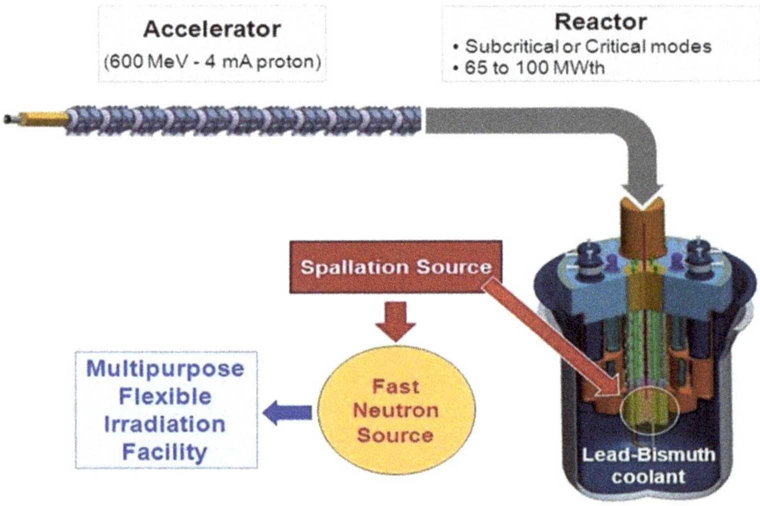

Figure 2.8. Schematic view of MYRRHA (Sornette *et al* 2018, with permission of Springer). Neutrons generated by spallation when the accelerated protons hit a lead–bismuth target are necessary to sustain the chain reaction in the reactor.

Uranium breeding chain: $^{238}_{92}\mathrm{U} + ^{1}_{0}\mathrm{n} \longrightarrow ^{239}_{92}\mathrm{U} \xrightarrow[23.5\ \mathrm{min}]{\beta^-} ^{239}_{93}\mathrm{Np} \xrightarrow[2.3565\ \mathrm{d}]{\beta^-} ^{239}_{94}\mathrm{Pu}$

Thorium breeding chain: $\overset{\text{neutron}}{\mathrm{n}} + ^{232}_{90}\mathrm{Th} \longrightarrow ^{233}_{90}\mathrm{Th} \xrightarrow{\beta^-} ^{233}_{91}\mathrm{Pa} \xrightarrow{\beta^-} \overset{\text{fuel}}{^{233}_{92}\mathrm{U}}$

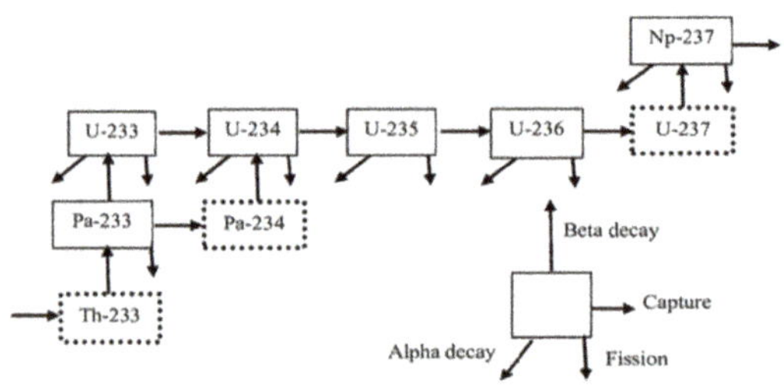

Figure 2.9. Comparison of the breeding chains of U-238 and Th-232 (Reproduced with permission by the IAEA from Tulsidas 2019).

capture and transmutation into higher isotopes (IAEA 2005). These isotopes decay into subsequent stable elements, eventually synthesizing the minor actinides, see figure 2.9.

The thorium cycle is very different: Th-232 is fertile, and upon neutron capture, it transmutes into protactinium Pa-233, which decays into U-233. U-233 has a very high fissile cross-section and only one-twelfth of the atoms will transmute into the fertile isotope U-234. Additionally, the Th-232 breeding chain is much lower on the scale compared to uranium; even successful neutron captures that fail to undergo fission will lead to the subsequent formation of fissile elements: U-235, Pu-239, and Pu-241. Therefore, the likelihood of the creation of transuranic waste (minor actinides) is much lower compared to that of the uranium cycle (by up to two orders of magnitude, Moir and Teller 2005). An additional factor in favor of thorium is its ability to undergo thermal breeding while maintaining a high burnup rate (60 GWd/tHM) (IAEA 2005), which is desirable for fuel utilization.

These characteristics, along with its high abundance in nature, make thorium a suitable alternative to uranium, which is why many of the proposed next-generation reactor designs aim to utilize thorium as their main fuel. The challenges that remain are the high radiotoxicity of certain isotopes in the thorium chain (Pa-231), the contamination of fissile U-233 by U-232 (which has a low fission chance of 50% upon neutron capture), and the lack of real-world operational experience. The usage of breeding blankets and the ability to easily separate Pa-233 in molten salt reactors prompt proliferation concerns, although thorium is generally seen as more proliferation resistant than uranium, mainly due to the U-232 contamination.

2.5 Conclusions

It is commonly agreed that the assessment of energy technologies has to be sufficiently comprehensive and holistic, rather than limited to single factors of specific interest. Such an assessment has to take the whole life cycle of a process or product into account, including all upstream and downstream processes. This kind of standard 'cradle-to-grave' evaluation method has also been applied to nuclear energy and examined nuclear fuel cycles in the context of other power-generation technologies. Among the merits of nuclear energy are low GHG emissions and low materials flows, including fuel flows and waste amounts, due to its incomparably high energy density; on the negative side are the potential for extremely rare but high-consequence accidents in nuclear power plants and long-lived radioactive waste that demands appropriately long stewardship times. The number of long-lived radionuclides and the characteristics of the material requiring deep geological disposal depend on the fuel cycle used.

According to the NEA (2021), fuel cycle options have been reduced to three: the open-cycle, mono-recycle, and multi-recycle options, which can be enhanced by the P&T of minor actinides. Some challenges and opportunities are shared by all the fuel cycle options, e.g., all require deep geological disposal, whereas requirements differ significantly. The significant differences among the fuel cycle options include the amount of uranium required at the front end and the volume of HLW at the back end. The decay heat load and the radiotoxicity of the material requiring deep geological disposal is significantly less for the multi-recycle option—in the case of enhanced recycling with P&T, the decay time to the natural uranium level shrinks to roughly 400 years, compared to 250 000 years for spent fuel in the open-cycle option.

It is important to note that effective multi-recycle options need fast reactors—a major technical challenge. Innovative to 'exotic' designs, which include the use of spent LWR fuel and actinide burners or the use of liquid metals or molten salts (with dispersed fuel) as coolants instead of water, are under development in many countries and have been introduced. The deployment times given vary from five to ten years, which appears to be rather optimistic when development and operation at a commercial scale are considered, and are associated with large uncertainties. Other ways of increasing fuel efficiency and relieving waste management issues were also introduced, in particular: striving for higher burnups, replacing uranium fuel by thorium, and solving the key issues of the transmutation of minor actinides. Technology readiness levels vary from high to low. Nevertheless, the period of time required to implement any new fuel cycle option is extremely long and should not put the work of finding solutions for deep geological disposals on hold.

Bibliography

Buongiorno J, Corradini M, John P and Petti D 2018 *The Future of Nuclear Energy in a Carbon-Constrained World* (Cambridge, MA: Massachusetts Institute of Technology) p 26

Debes M 2001 Increased fuel burn-up and fuel cycle equilibrium *9th Int. Conf. on Nuclear Engineering* vol 33 *(Nice Acropolis, France, 8-12 April 2001)* (Vienna: IAEA) Report INIS-FR--749

Garzarolli F, Manzel R and Stehle H 1979 KWU fuel performance with emphasis on high burnup *ANS Topical Meeting on LWR Fuel Performance (Portland, Oregon)* Report DOE/ET/34007--1

Hao C *et al* 2020 Sensitivity and uncertainty analysis of the maximum fuel temperature under accident condition of HTR-PM *Sci. Technol. Nucl. Installations* **2020** 9235783

Hirschberg S and Burgherr P 2015 Sustainability assessment for energy technologies *Handbook of Clean Energy Systems* vol 6, ed S K Chou, Y M Wei and J Yan (Hoboken, NJ: Wiley)

Hitachi G E, Hitachi G E and PRISM 2018 Selected for U.S. Department of Energy's Versatile Test Reactor Program https://ge.com/news/press-releases/ge-hitachi-and-prism-selected-us-department-energys-versatile-test-reactor-program

IAEA (International Atomic Energy Agency) 2002 *Technical and Economic Limits to Fuel Burnup Extension* IAEA-TECDOC-1299

IAEA (International Atomic Energy Agency) 2005 *Thorium Fuel Cycle — Potential Benefits and Challenges* IAEA-TECDOC-1450

IAEA (International Atomic Energy Agency) 2019 *World Thorium Occurrences, Deposits and Resources* IAEA-TECDOC-1877

IEA (International Energy Agency) 2019 *World Energy Outlook 2019* https://iea.org/reports/world-energy-outlook-2019

Kröger W, Sornette D and Ayoub A 2020 Towards safer and more sustainable ways for exploiting nuclear power *World J. Nucl. Sci. Tech.* **10** 91–115

Moir R W and Teller E 2005 Thorium-fueled underground power plant based on molten salt technology *Nucl. Technol.* **151** 334–40

Motta A T, Couet A and Comstock R J 2015 Corrosion of zirconium alloys used for nuclear fuel cladding *Ann. Rev. Mater. Res.* **2015** 311–43

MYRRHA Project 2020 *MYRRHA Phased Implementation* www.myrrha.be/myrrha-project [accessed 20 September 2021]

NEA (Nuclear Energy Agency) 2011 *Potential Benefits and Impacts of Advanced Nuclear Fuel Cycles with Actinide Partitioning and Transmutation.* (Paris: OECD Publishing) NEA No. 6894

NEA (Nuclear Energy Agency) 2018 *State-of-the-Art Report on Light Water Reactor Accident-Tolerant Fuels* NEA No. 7317

NEA (Nuclear Energy Agency) 2021 *Strategies and Considerations for the Back End of the Fuel Cycle* NEA No. 7469

NEA (Nuclear Energy Agency) and IAEA (Int. Atomic Energy Agency) 2018 *Uranium 2018: Resources, Production and Demand* NEA No. 7413

NEEDS (New Energy Externalities Developments for Sustainability) 2016 *NEEDS Documents* www.needs-project.org [accessed 20 September 2021]

NEI (Nuclear Engineering International) 2019 *Titan-2 Contracted to Build Russia's Brest 300 Reactor* www.neimagazine.com/news/newstitan-2-contracted-to-build-russias-brest-300-reactor-7527297 [accessed 20 September 2021]

Seaborg Technologies 2015 *Seaborg Waste burner. Molten Salt Reactor*. White paper SEAB-WP-2015-001 http://citeseerx.ist.psu.edu/viewdoc/download?doi=10.1.1.712.5069&rep=rep1&type=pdf

Sornette D, Kröger W and Wheatley S 2018 *New Ways and Needs for Exploiting Nuclear Energy* (Berlin: Springer)

Tulsidas H 2019 *Nuclear Fuel Cycle Simulation System: Improvements and Applications* IAEA-TECDOC-1864

WNA (World Nuclear Association) 2020 *Uranium Markets* June 2020 www.world-nuclear.org/information-library/nuclear-fuel-cycle/uranium-resources/uranium-markets.aspx [accessed 20 September 2021]

Zhang Z *et al* 2016 The Shandong Shidao Bay 200 MWe high-temperature gas-cooled reactor pebble-bed module (HTR-PM) demonstration power plant: an engineering and technological innovation *Engineering* **2** 112–18

IOP Publishing

Nuclear Waste
Management, disposal and governance
Klaus-Jürgen Röhlig

Chapter 3

Other waste origins (industry, medicine, research): sources and resulting waste types

Michael I Ojovan

3.1 Radioactive waste types

The International Atomic Energy Agency (IAEA) classifies radioactive waste into six categories according to the activity and half-life of waste radionuclides (IAEA 2009): (a) exempt waste (EW); (b) very short-lived waste (VSLW); (c) very low-level waste (VLLW); (d) low-level waste (LLW); (e) intermediate-level waste (ILW) and (f) high-level waste (HLW). Radioactive wastes are either *controlled* or *uncontrolled.* Controlled wastes are largely a product of the nuclear fuel cycle (NFC) used to generate electricity for civil use. Controlled non-NFC wastes include those from various applications of radionuclides in research, medicine, and industry, including spent or disused sealed radioactive sources (DSRS) of isotopes used in medical applications. Uncontrolled wastes arise when unexpected events occur or where the level of care was not taken that would be expected today (Lee *et al* 2013). Radioactive waste is also produced by technological processes that concentrate naturally occurring radionuclides in materials such as coal, oil, and water and during the utilisation of nuclear energy in medicine, research, geological surveys, agriculture, and food preservation (Ojovan *et al* 2019). Non-NFC institutions include many industrial users of radionuclides, research organisations such as nuclear research centres (some of which operate research reactors), medical, and other small users. The radioactive wastes from non-NFC institutions typically occur in the form of medical residues that are basically contaminated by very short-lived radionuclides, solid wastes from laboratories and research reactors, contaminated soil and debris, washing and decontamination liquids, aqueous solutions from the regeneration of ion exchangers, and spent sorbents (table 3.1).

Typically, the non-NFC institutions generate LLW; however, they can also produce highly radioactive waste such as DSRS from medical applications and

doi:10.1088/978-0-7503-3095-4ch3

Table 3.1. Waste types from radionuclide applications. Reproduced with the permission of Elsevier from Ojovan *et al* (2019).

Solid	Aqueous	Organic liquids	Highly radioactive
Metallic scrap, brickwork, sorbents including ion-exchange resins, glassware, filters, cardboard, plastics, paper, swabs, tissues, protective clothes, gloves.	Effluents from laboratories, hot cells, fuel storage pools, sumps, decontamination, rinsing water, mining–milling raffinates.	Pump oils, scintillation liquids, extraction solvents such as tributyl phosphate (TBP), kerosene and amines.	Sealed radioactive sources, radium needles, reactor-irradiated materials.

defectoscopy. The IAEA gives specific guidance on the management of radioactive non-NFC waste and DSRS (IAEA 2005, 2019).

3.2 Radioactive materials with natural radionuclides

In the mining industries and at industrial processing sites (mining and processing plants, oil and gas processing, fertilizer production, etc.), large amounts of waste are generated that are contaminated by natural radionuclides contained in natural raw materials. Materials which contain natural radionuclides are designated as naturally occurring radioactive materials (NORM). The most common radionuclides in NORM are C-14, K-40, isotopes of uranium, thorium, etc. Some technological processes, such as ore melting or coal incineration result in the concentration of radionuclides in industrial residues and therefore in the generation of radioactive waste. Waste that contains NORM is known as NORM waste. NORM waste, which typically contains radionuclides of the uranium and thorium decay chains, is characterized by very large volumes. In view of its potential hazard, NORM waste handling requires compliance with the relevant requirements established by regulatory authorities, although these requirements are usually less stringent than those imposed on artificial radioactive materials. In addition, NORM waste often contains other toxic substances, such as heavy metals, and for this reason, both radiological and non-radiological aspects are important for its management.

3.3 Radioactive waste from medical applications

The use of radioactive sources of radiation and radioactive materials in medicine enables the pursuit of diagnostic and therapeutic goals. The main applications of radionuclides in medicine are in radio-immunoassays, radiopharmaceuticals, diagnostic procedures, and radiotherapy. The radionuclides used in hospitals for medical diagnostic procedures and treatments are very short-lived, and the waste generated is usually stored for decay prior to further treatment as non-radioactive waste. Positron emission tomography (PET), for example, incorporates cyclotron-generated C-11 which has a 20-minute half-life, or F-18 with a 110-minute half-life in a molecule such as sugar, which is intravenously administered to the patient and is detected during its circulation around the body. However, some radionuclides used in medical applications have longer half-lives, including Co-57 (271.7 days), which is used in clinical measurements, tritium (12.3 years), and C-14 (5730 years), which is used in radiolabelling (Ojovan *et al* 2019). Radioanalysis procedures generally use kits containing radiation sources that have activity levels of several kBq; they are most often based on I-125 or Tc-99m. After each individual analysis and after the expiration date of the kit, the radioactive material and any contaminated items should be treated as radioactive waste. The type and amount of oral radio-pharmaceutical preparation (*in vivo*) taken by a patient is determined by the organ to be examined or treated. For example, the most-used radionuclide, Tc-99m with a half-life of 6 h, is usually obtained from commercially available generators containing Mo-99. Since the half-life of Mo-99 is 66 h, the generators should be replaced approximately every week. Flasks, syringes, tampons, etc. used for the preparation of Tc-99m use are classified as VSLW. Whether the urine and excrement of patients, bedding, etc. is radioactive waste depends on the characteristics of the internal radiopharmaceuticals. Basically, radiopharmaceuticals contain limited amounts of short-lived radionuclides that result in waste that typically requires the same handling as ordinary household waste. However, this is not a generic situation and each case must be considered separately, as in the case, for example, of the more active waste arising from the use of radiopharmaceuticals based on I-131, P-32, Y-90, and Sm-153. The medical applications of radionuclides such as those used for bone densitometry, manual brachytherapy, and whole-blood irradiation may not only use small quantities of unsealed sources and liquid solutions, but also of highly radioactive sealed radioactive sources (SRSs) housed in shielded assemblies.

Basically, the wastes produced by medical institutions are solid waste (surgical instruments, dressings, parts of dismantled irradiation installations, and other products made of various materials), the most active of which are SRSs that have expended their resource and become spent or disused. The total activity of an SRS depends on its purpose and the nature of the radiation. In medical practice, SRSs are used for external irradiation of tumours (for the so-called teletherapy), for intratissue and intraband irradiation (brachytherapy) as both permanent and temporary implants, as well as for sterilizing blood samples, syringes, and surgical instruments. Teletherapy uses large sources, often based on Co-60, possibly Cs-137, with an activity of several hundred TBq; for brachytherapy—sources based on Ir-192, sometimes Co-60 and Cs-137 (whose activity is tens of TBq) (Ojovan *et al* 2019). SRSs whose activity has decreased to levels unacceptable for their further use for their direct purpose due to natural decay and damaged sources are considered radioactive waste. DSRSs are extremely hazardous, as they may contain large quantities of radionuclides.

3.4 Radioactive waste from research activities

Almost every radionuclide of any activity can find an application in research activities, as a result of which, solid, liquid, and biological radioactive wastes are formed.

The liquid radioactive wastes created by research activities are very diverse. Here are some examples: in small nuclear centres, liquid waste can be contaminated by both short-lived (half-life below 30 y) and long-lived (half-life above 30 y) radionuclides such as C-14 and H-3. As a result of radiochemical neutron activation analysis carried out in research reactors, aqueous waste is formed that can have an extremely varied chemical composition and contain mainly relatively short-lived radionuclides. Part of the liquid radioactive waste generated by metabolic studies can be biologically hazardous water waste. Liquid aqueous waste arising from the decontamination of equipment may contain large amounts of the complexing agents used to dissolve radioactive contamination. Nuclear research centres are the main originators of organic liquid waste due to their use of the organic solvent TBP for the extraction of U and Pu. For the extraction of heavy metals, other organic compounds are sometimes used, for example, amino compounds, but these are used in smaller quantities than TBP. Due to the variety of research programs, the nature and quantity of generated radioactive waste may be specific to each research project. We have listed below some of the areas of scientific research in which radionuclides are used and radioactive waste is generated.

The manufacture and labelling of formulations in biological research. The range of radionuclides used, and their quantities are usually small (H-3, I-125, C-14, or P-32), and the activity of the generated waste corresponds to several MBq for the radionuclide used. However, the total and specific activity of wastes from the radionuclide labelling process can be quite high due to the large amounts of these wastes.

Clinical and pharmaceutical products, metabolic, toxicological, or environmental problems. These studies investigate a wide range of formulations of drugs, pesticides, fertilizers, and minerals. The most-used radionuclides are easily incorporated into complex molecules and are easy to trace, e.g. C-14 and H-3. Studies can be carried out on animals; in this case, such waste appears as radioactive excrement, corpses, bedding, etc. Animal corpses can additionally pose both biological and chemical hazards.

Nuclear fuel cycle. This research is usually carried out in laboratories using small amounts of fissile material (U, Pu) and fission products (mainly Cs-137 and Sr-90), depending on the task. The wastes generated include solids and liquids that contain used fissionable material and fission products.

Physics, chemistry, and biology. These studies using radionuclides can include non-destructive testing, solid-state physics, and the development of new materials, etc. The spectrum of radioisotopes used is very broad.

Research reactors. Many institutions worldwide have installed small reactors in support of research and development programmes. Research reactors provide a wide range of training, research, commercial, and nuclear power programme support functions; these reactors are generally not used for power generation. Their output (neutron beams) is used for the non-destructive testing and analysis of materials, the production of radioisotopes, research, and public outreach and education. For example, the UK has had more than 30 research reactors since the

1950s, but like all countries, it has shut down and decommissioned almost all of them (Lee *et al* 2013). Solid radioactive waste from research activities may include activated materials, such as shielding elements, isotope production containers, and materials that have been tested in research reactors. The activity of such wastes, determined by the presence of Co-60 and other activated impurities in the steel, is a function of irradiation. A separate group of wastes from research activities (like any industry that uses radioactive materials) consists of spent and damaged DSRSs.

3.5 Radioactive waste from industrial uses of radioactive materials

SRSs are widely used in various industries: for non-destructive analysis (radiography and measurement of various production parameters), the sterilization of food and other products, technological process control and equipment calibration, etc. SRS activities depend on the purpose of the related applications and the nature of the source radiation. The most potentially dangerous are powerful irradiation installations in which food and medical products are sterilized and chemicals are polymerized. These relatively few installations contain sources based on Co-60 and Cs-137 of very high activity (0.2 to 600 PBq). Widespread industrial radiographic sources also have a high degree of potential hazard. Thus, during the construction and maintenance of petrochemical facilities, sources with activity levels of up to 7 TBq based on Cs-137, Ir-192, or Co-60 are often used to test welds. The sources used for calibration are less active, and the range of radionuclides of these sources and their activities is very wide. The same can be said for logging devices and sources used in the industry to measure the level, thickness, density, moisture content, or the presence of any material during its extraction, manufacture, or processing. Industrial sensors can contain relatively small amounts of radioactive material.

SRSs fall under the category of radioactive waste when their activity, as a result of natural decay, decreases to a level that makes them unsuitable for their original purposes. In addition, industrial DSRS are considered as waste when the devices (equipment) containing them are outdated, worn out, or for some other reason no longer necessary, or mandatory radiation monitoring has revealed a radioactivity leak.

Relatively large amounts of solid, liquid, and gaseous waste are formed when the source enters the scrap metal sent for remelting. In addition to contaminating the production process, radioactive material can be included in new products.

An example of the industrial use of open sources as a radioactive tag is the assessment of wear and corrosion of key components of an installation or machine tools, for example, engine parts, the inner surfaces of furnaces, etc. Other examples of the use of a radioactive tag are the monitoring of wastewater treatment and studies of the movement of groundwater, the dispersion and dilution of water, or gaseous discharges. In most cases, small labels of short-lived radionuclides are used, which does not lead to the formation of radioactive waste.

3.6 Radioisotope production

Particle accelerators and research nuclear reactors are used to produce radioisotopes. Radioisotopes are produced on targets and in capsules, which are removed from installations after irradiation for processing and purification. As a result, small volumes of liquid radioactive waste of relatively high activity and larger volumes of low-level waste are formed. Research institutions may also use some of the radionuclides produced in their own research programmes. In the industrial production of radioisotopes, liquid waste is formed mainly from the dissolution of targets; the volume and pollution of this waste are low. Depending on the chemical processes, the wastewater solutions produced as waste can be chemically reactive.

3.7 Disused sealed radioactive sources

Sealed radioactive sources (SRS) are widely available and are used extensively in a broad range of applications, including medical, industrial, and research and education. They are generally small (typically a few centimetres in size) and the radionuclides are typically enclosed in capsules made, with very few exceptions, of stainless steel, titanium, platinum, or other inert metals. Source capsules are airtight and durable. The radionuclides may also be closely bonded to a substratum. The recommended useful life of most sealed sources is 5–15 years, but they represent a potential radiation hazard long after the devices containing them have been decommissioned. In most cases, the source capsule will be undamaged at the time it is collected as a DSRS for long-term storage or disposal. However, DSRS integrity cannot be taken for granted either before or after collection, nor can the longevity of the source capsule be guaranteed after disposal. Source SRS activity levels vary over a considerable range from low activity to very highly active, with half-lives varying from some days to thousands of years (figure 3.1).

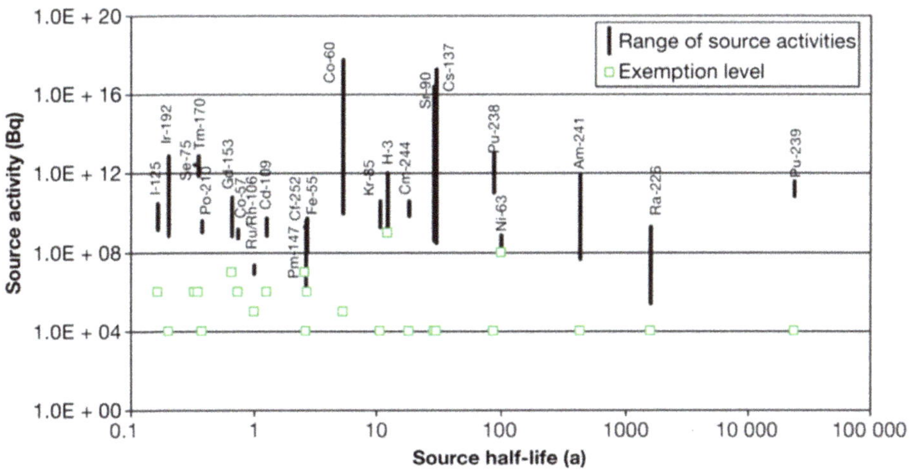

Figure 3.1. Radionuclides, half-lives, and typical activity levels of SRSs. Reproduced with permission of the IAEA from IAEA (2005).

3.8 Radioactive waste arising from radiation accidents

As a result of accidents and during their cleanup, wastes of various volumes, states of aggregation, levels of activity, and other properties arise, depending on the scale of the accident and the radioactive materials involved in them. Such radioactive wastes can take the form of contaminated soil, various pieces of equipment, building materials and structures, contaminated vehicles, water from decontamination, etc.

Bibliography

IAEA 2005 *Disposal Options for Disused Radioactive Sources* Technical Report No. 436

IAEA 2009 *Classification of Radioactive Waste* General Safety Guide No. GSG-1 https://www-pub.iaea.org/MTCD/Publications/PDF/Pub1419_web.pdf

IAEA 2019 *Predisposal Management of Radioactive Waste from the Use of Radioactive Material in Medicine, Industry, Agriculture, Research and Education* Specific Safety Guide SSG-45

Lee W E, Ojovan M I and Jantzen C M 2013 *Radioactive Waste Management and Contaminated Site Clean-Up* (Cambridge: Woodhead)

Ojovan M I, Lee W E and Kalmykov S N 2019 *An Introduction to Nuclear Waste Immobilisation* 3rd edn (Amsterdam: Elsevier)

IOP Publishing

Nuclear Waste
Management, disposal and governance
Klaus-Jürgen Röhlig

Chapter 4

Waste management policy and strategy

Klaus-Jürgen Röhlig

4.1 Policies and strategies: what should they contain?

As for all hazardous materials, the use of well-planned management is obviously advisable in the case of radioactive waste. In other words, a management strategy (or several strategies for the various waste types) should be developed and implemented. More precisely, the anticipated fate of each waste type, as well as the route towards this fate, should be described systematically in a transparent way. This, of course, means that the first step in developing a management strategy should be concise and comprehensive stocktaking: what waste types and amounts are to be managed, and what arrivals can be expected in the future? Such a strategy needs to account for political, legal, financial, and technological possibilities and boundary conditions and foresee measures for implementation that address organisational, financial, and human resources aspects.

A strategy, though, is not self-sufficient—it should serve purposes or goals. According to IAEA terminology and guidance (see below), the latter are laid down in a so-called policy containing e.g. objectives for the protection of humans and the environment, political goals, or principles of transparency and public participation.

However, given that radioactive waste management strategies will typically address processes that take decades, it is obvious that they will not be implemented exactly as initially planned, but will undergo changes that account for changing boundary conditions; there might, for example, be surprises in the form of novel wastes, technological developments, or changes in the aspirations related to civil society engagement. Therefore, strategies need to be flexible and account for uncertainties.

The terms 'policy' and 'strategy' and their relationship are defined by the International Atomic Energy Agency (IAEA), as follows:

> **Policy** is a set of established goals or requirements for the safe management of spent fuel and radioactive waste; it normally defines national roles and

doi:10.1088/978-0-7503-3095-4ch4

responsibilities. As such, policy is mainly established by the national government; policy may also be codified in the national legislative system.

Strategy is the means for achieving the goals and requirements set out in the national policy for the safe management of spent fuel and radioactive waste. Strategy is normally established by the relevant waste owner or operator, either a governmental agency or a private entity. The national policy may be elaborated in several different strategies. The individual strategies may address different types of waste (e.g. reactor waste, decommissioning waste, institutional waste, etc.) or waste belonging to different owners.

The line separating policy from strategy is not always sharp and sometimes it is unclear whether an issue should be addressed by a policy or a strategy (IAEA 2009a, p 5).

More precisely, strategies are meant to define 'the ways in which the various types of radioactive waste (including spent fuel if declared waste) in the country or site will be managed during all phases of the radioactive waste life cycle (from cradle to grave)' (IAEA 2009a, p 8). In the opinion of the author, this includes not only the definition and planning of individual management steps, but also the anticipation of problems and obstacles as well as the means to overcome them.

'The organization responsible for radioactive waste management implements a strategy by selecting one or more technologies available to manage the waste in accordance with national policy stipulations or legal regulations. Ideally, the strategy should be determined before the system is put in place, though in practice this may be not the case.' (IAEA 2017, p 1).

Boxes 4.1–4.3 give brief and cursory descriptions of some national examples of strategies (or the lack thereof). Some of the details and terms mentioned might only be understandable after having read the relevant chapters of this book, but nevertheless, the content might give an impression of the variation in approach and the problems that might occur when defining and implementing strategies.

Generally, extensive descriptions of national strategies are provided using the reporting system of the Joint Convention on the Safety of Spent Fuel Management and on the Safety of Radioactive Waste Management (IAEA 1997, 2018a). Shorter, but possibly out-of-date information is given in the country profiles and reports of the Nuclear Energy Agency (NEA 2020a).

In IAEA guidance, various options for assigning responsibilities for the definition and implementation of strategies are named. Strategies might be formulated at the national level or by companies (e.g. waste owners); they might address radioactive waste in general, or only particular waste types (IAEA 2009a, p 21). The situation also varies from country to country since, in some countries, the organisations responsible for radioactive waste management are owned by the state, but in others, they are owned by waste producers. According to the EU Directive on the

Box 4.1. ***Strategies for managing radioactive waste in brief: the Swedish example* (SKB 2021).**

In Sweden, there are no plans to reprocess Spent Nuclear Fuel (SNF). There are plans to dispose of the SNF in a designated repository at Forsmark, the licensing procedure for which is ongoing. Currently, SNF is stored centrally underground in a wet storage facility ('Clab'—in Swedish, 'Centralt mellanlager för använt kärnbränsle' = Central intermediate storage for spent nuclear fuel) at Oskarshamn, next to which conditioning for disposal in an 'encapsulation plant' is foreseen (both facilities together will then be called 'Clink' for 'Centralt mellanlager och anläggning för inkapsling av använt kärnbränsle' = Central interim storage and encapsulation plant for spent nuclear fuel). SNF transport from the nuclear power plant to Clab/Clink takes place by sea; the same is planned for the transport of the encapsulated (conditioned) SNF to the repository.

The destination for low- and intermediate-level waste (LILW) from the operation and decommissioning of nuclear facilities and from other sources (industry, medicine, research) depends on its lifespan: short-lived waste is taken to the SFR facility at Forsmark (slutförvaret för kortlivat radioaktivt avfall = final repository for short-lived radioactive waste); for longer-lived waste (currently in interim storage), a dedicated facility is foreseen (slutförvar för långlivat avfall, SFL), the development of which is less advanced. Very low-level waste which cannot be cleared goes to surface repositories.

management of spent fuel and radioactive waste, the main responsibility lies with the member states, which are required to lay down important policy and strategy elements in national programmes:

'Member States should establish national programmes to ensure the transposition of political decisions into clear provisions for the timely implementation of all steps of spent fuel and radioactive waste management from generation to disposal. It should be possible for such national programmes to be in the form of a single reference document or a set of documents.' (EU 2021, preamble)

Given the heterogeneity in both terminology and responsibilities, in the following, the elements of policies and strategies are addressed without drawing a sharp line between the two. According to IAEA guidance (IAEA 2009a, p 12 ff.) and EU requirements (EU 2021, Articles 12 and 10), policies and strategies should comprise at least the following:

- Overall objectives, particularly with respect to safety and security
- Descriptions of waste inventories and prognoses of future waste arisings
- Assignment of responsibilities
- Concepts, technical procedures, and measures 'from cradle to grave,' including 'end points' (e.g. disposal) and the post-closure period of disposal facilities
- Time schedules and implementation plans
- Supervisory mechanisms and the indicators used to monitor progress
- Cost estimates and funding schemes
- Plans for research, development, and demonstration
- Information, transparency, and participation issues
- Agreements with other states

Box 4.2. ***Strategies for managing radioactive waste in brief: the Canadian example*** **(NRCAN 2015, OPG 2021).**

In Canada, no reprocessing of SNF is foreseen. For its management and disposal, a strategy called 'Adaptive Phased Management' has been developed, which is now in its early stages. This strategy is directed at geologic disposal and is flexible about the route taken toward that goal; it includes the possibility of accounting for stakeholders' requests, societal changes, and technological developments (see below and part I, chapter 7 'Disposal and other conceivable strategy 'endpoints' for different types of waste'.

Canada distinguishes LILW (in Canadian documents: 'L&ILRW') according to its origins: 'ongoing waste' is generated by operational nuclear facilities, 'Canada's Nuclear Legacy Liabilities' were produced by activities during the Cold War and the early developments of nuclear technologies, and 'historic low-level radioactive waste [...] that was managed in the past in a manner no longer acceptable, for which the current owner cannot be reasonably held responsible and for which the Government of Canada has accepted responsibility for long-term management'.

The ongoing waste is currently stored by its producers (nuclear power plants). One of them, Ontario Power Generation (OPG), developed a geologic disposal project at the Bruce nuclear site in Kincardine (Ontario) which was supported by local authorities; this, however, failed due to a veto by the Saugeen Ojibway Nation. Currently, OPG is exploring options for LILW long-term management, as are other waste owners, some of whom also take care of waste from small producers. In other words, in contrast to the case in Sweden, there is no centralised management of this waste.

For the legacy waste, a specific programme called the 'Nuclear Legacy Liabilities Program (NLLP)' is in force, which aims to develop a long-term management strategy and, at the same time, undertake first steps, such as waste characterisation, decontamination, the recovery of buried waste, and other activities that aim to reduce environmental and financial risks and prepare for the next management steps.

For the historic waste (mostly contaminated soils from clean-ups), disposal in above-ground mounds is planned or, in one case, already implemented.

Of course, all these elements are embedded in a much larger frame: overall objectives are determined by national legislation and regulation as well as by international obligations and treaties. As well as the issues of safety and security mentioned above, environmental protection and fairness considerations are of the utmost importance when defining objectives; see part II, chapter 13 'Ethical aspects of high-level nuclear waste management'.

Waste inventories and future requirements depend on the use of fissile and radioactive materials in the country under consideration and in particular on its nuclear energy programme (or lack thereof).

Responsibility assignment has to fit into the overall legal and administrative framework and serve the requirements of good governance. Moreover, the independence of regulatory and licensing authorities has to be ensured (EU 2021, Article 6, IAEA 1997, Articles 2, 19, 20; see also part II, chapter 10 'Legal and regulatory issues').

Box 4.3. ***Direct disposal or reprocessing? Central or decentralized storage? 'Muddling through'*[1] in Germany facing political, economic, and societal uncertainties (Dutton *et al* 2004, Endlagerkommission 2016).**

Before 1994, the reprocessing of SNF was mandatory in Germany. The idea was to develop a partially closed fuel cycle (see part I, chapter 2 'Nuclear fuel from cradle to grave: existing variants and future options for the fuel cycle and resulting waste types') that included fast breeder reactor(s), reprocessing in Germany, and a nuclear waste management center ('Nukleares Entsorgungszentrum NEZ') at Gorleben (Lower Saxony) with storage, reprocessing, and conditioning facilities as well as final geologic disposal of all types of radioactive waste in the Gorleben salt dome, which was investigated for this purpose.

The growing scepticism concerning nuclear activities in parts of society and the increasing resistance that resulted, *inter alia*, in the failure of various projects (including the reprocessing facility at Wackersdorf and the fast breeder reactor at Kalkar), lead to the first changes: from 1994 onward, both direct SNF disposal and reprocessing were allowed. SNF reprocessing took place abroad (in France and the UK), and a commitment was made to take back the resulting waste, particularly vitrified high-level waste (HLW). This waste, as well as SNF to be disposed of directly, was to be stored centrally at Gorleben; a second central SNF storage facility was implemented at Ahaus (North Rhine-Westphalia). However, Gorleben became a focal point of the antinuclear movement, particularly on the occasion of the HLW transports from reprocessing for storage at Gorleben. Further SNF storage facilities were implemented at Jülich (North Rhine-Westphalia, for SNF from the experimental reactor located there) and Lubmin (near Greifswald, Mecklenburg-Hither Pomerania, for SNF from East German reactors which were to be decommissioned after German unification in 1991).

When the Federal Government came to an agreement in 2000 with the electricity providers about a nuclear power phase-out, this agreement was linked to a decision to abstain from central SNF storage. Instead, decentralised storage was to be implemented at reactor sites. SNF transport to reprocessing facilities in France and the UK became banned from 2005, but HLW from reprocessing abroad still went to Gorleben. The investigation of the Gorleben salt dome was interrupted.

The partial revision of the nuclear phase-out in 2010, which included considerable lifetime and production allowance extensions for the nuclear reactors was accompanied by a resumption of the site investigation at Gorleben. However, the situation changed again after the Tōhoku earthquake and tsunami and the resulting nuclear accidents at Fukushima in 2011: the extensions were revoked, and the phase-out was now (as opposed to the situation in 2000) backed by a broad political consensus which also allowed a novel approach to the until-then highly controversial issue/question of SNF/HLW disposal. A site selection process was started (see part I, chapter 8 'Geologic ('deep') disposal of high-level and other long-lived waste: host rocks, concepts, current international status' and part II, chapters 15 'Geology, engineering, and society: repository siting as a socio-technical problem' and 16 'The German case

[1] Muddling through is a term introduced by Ortwin Renn that describes a strategic approach that relies 'on minimum consensus [...], which emerges from the political opinion process. The only options to be considered legitimate are those that induce the least amount of opposition within society.' (Streffer *et al* 2011, p 84).

for dealing with high-level radioactive waste: taking a socio-technical approach to address a socio-technical problem—chances and risks'). In order to avoid further tensions related to Gorleben, the remaining HLW transports from reprocessing were directed to four of the decentralised storage facilities mentioned above.

As a consequence of the often-changing societal, political, and legal boundary conditions, there are now 17 SNF storage sites in Germany, five of which have also been, are also, or will also be accepting HLW from reprocessing. While many of these were initially connected to nuclear power plants and almost all were owned by subsidiaries of the energy providers, since 2019, the ownership of all of them has been in the hands of companies owned by the Federation ('Bundesgesellschaft für Zwischenlagerung mbH BGZ' and 'Entsorgungswerk für Nuklearanlagen GmbH EWN'). Even if site selection for HLW/SNF disposal takes place, as planned, in 2031, the lifetimes according to the licenses of the facilities will not be sufficient for final disposal to be implemented.

A clear strategy to address the situation is still missing; according to the nuclear waste management programme (BMU 2015), the first ideas involve a 'reception storage facility' at the disposal site to be determined. For research activities addressing this question, see part I, chapter 6 'Storage and transport'.

Concepts and schedules have to account for technological and financial capabilities and capacities (including perspectives for the future) as well as for boundary conditions with regard to governance and participation.

Funding schemes should fulfil the obligations of the 'polluter pays' principle and must, at the same time, account for the diversity of waste producers and their ownership and maximize funding security, even in the long term. The issue of funding is politically sensitive and subject to uncertainties related to both the evolution of waste management programmes and the financial and economic situation over the timeframes of concern. Funding considerations often go beyond the scope of radioactive waste management by also including the costly issue of decommissioning nuclear facilities, in particular, nuclear power plants. In principle, the 'polluter pays' principle should apply. However, costs are strongly influenced by strategic and policy choices, which are often, at least in large part, not defined by the 'polluters' (i.e. the waste producers). NEA (2021a) provides an overview of the funding approaches applied in a number of countries and advocates a 'circular approach' in which 'the interdependent elements of funding arrangements continuously adapt to the extent that new information on costs, social preferences, policy objectives, lifetimes, or rates of return on capital become available.'

Research programmes have to fulfil the needs of radioactive waste management while accounting for existing research capacities. There is also the question of which research, development, and demonstration activities fall under which responsibility; of course, the organisations responsible for waste management have to ensure that their needs are addressed, but there are also regulatory requirements and those of basic research; the latter often fall under the responsibility of the state. A diversity of research activities has to be ensured in order to allow for the independence of regulators, out-of-the-box thinking, and the 'four-eyes' principle.

Information and participation should be coherent with the general principles, procedures, and political culture in the country under consideration. However, the specifics of activities related to nuclear power and the use of radioactivity might make it necessary to implement participation procedures that go beyond those applied in other contexts, especially when it comes to the siting of waste management, in particular disposal, facilities.

An important, but apparently sometimes ignored, point is that strategies (and even policies) need to be flexible in the presence of changing boundary conditions and internal evolutions. Such changes might be related to the perspectives of a nuclear programme (e.g. a phase-out decision as in Germany or, much earlier, in Italy) or to the approach to reprocessing (e.g. Belgium, Switzerland, Germany). Most notably, significant delays or even failures of disposal programmes sometimes force changes to the whole strategy (in particular, but not only, with respect to storage times).

Strategies tend to disregard this point of evolution and changes over time and instead use a 'linear' and inflexible approach and time schedule. In its first paragraph, the German Site Selection Act claims to initiate a process which is, amongst other things, self-questioning, learning, and reversible, but gives little leeway to such reversibility in its ensuing paragraphs (see the wording in German at http://www.gesetze-im-internet.de/standag_2017/, https://www.base.bund.de/EN/soa/siteselectionprocess/siteselectionprocess.html with a brief English excerpt [both accessed 5 March 2021]). In Canada, in contrast, the concept of 'Adaptive Phased Management (APM)' appears to aim at flexibility and the potential for adaption by means of 'both a technical method and a management system,' see https://www.nwmo.ca/en/Canadas-Plan/About-Adaptive-Phased-Management-APM [accessed 5 March 2021].

In the course of designing strategies, it seems advisable to develop schedules with appropriate timelines (EU 2021, Article 12), but also with decision points at which strategies might be adjusted or revised. Moreover, there is a need to be aware of the detrimental effects of potential path dependence, i.e. 'the tendency of institutions or technologies to become committed to develop in certain ways as a result of their structural properties or their beliefs and values', https://www.britannica.com/topic/path-dependence [accessed 5 March 2021].

The most comprehensive and recent overview of national policies and strategies can be found under the umbrella of the reporting system related to the Review Meetings of the 'Joint Convention on the Safety of Spent Fuel Management and on the Safety of Radioactive Waste Management Documents'; the most recent meeting was held in 2018 (IAEA 1997, 2018a), see part II, chapter 10 'Legal and regulatory issues.'

4.2 Technical elements of a strategy and decisions

4.2.1 Objectives of strategies and technical elements with which to address them

Obviously, the definition of policies and strategies requires that binding decisions are taken in order to address various, perhaps sometimes conflicting, objectives. Such objectives and the boundary conditions under which they are to be achieved are related to the following issues (compiled based on Dutton *et al* 2004, CoRWM 2006, IAEA 2009a, IAEA 2017, Röhlig *et al* 2019, NEA 2021b):

- The safety of staff and the general public, security, non-proliferation, radiation protection, environmental protection, environmental impact
- Ethical aspects (including inter- and intragenerational justice and responsibility)
- Technical and technological aspects (availability and perspectives of technologies vis-à-vis their potential to address objectives)
- Societal and political aspects (governance, political feasibility, acceptance)
- Costs and financial security
- Practical considerations (e.g. flexibility, minimising transport, …)
- Boundary conditions e.g. those arising from from legislation and regulation (which might address aspects of the points above) or from the nuclear energy programme

Note that these issues are interrelated (e.g. safety and environmental protection are ethical issues), apply at different levels (e.g. while safety can be considered an overarching principle, technology availability has to do with practicalities), might evolve over time (which is the case e.g. for safety standards or technology availability), can be actively influenced (e.g. technology availability), and might conflict with each other (e.g. safety at different times in the future). The way in which these issues are weighed and balanced against each other is also important.

For the particular case of strategies for the management of SNF ('fuel cycles'), the NEA (2021b, table 1) considers these aspects to be either challenges, opportunities, or risks.

A waste management strategy should cover the 'life journey' of the waste materials from the 'cradle', i.e. the waste's origin, to the 'grave', e.g. the ultimate fate of the materials. It is fair to say that decisions concerning the 'cradle' as well as the 'grave' have the highest impact on the 'life journey' of the materials in question.

When considering the 'cradle,' in the interest of sustainable development, one might consider replacing the notion of 'waste management' by 'resource management' or 'life-cycle management' for radioactive materials, as is already becoming apparent for other types of waste. Further elaborations can be found below in section 4.2.2 and particularly in part I, chapter 2 'Nuclear fuel from cradle to grave: existing variants and future options for the fuel cycle and resulting waste types.' The 'grave' is further addressed in 4.2.3 and particularly in part I, chapter 7 'Disposal and other conceivable strategy 'endpoints' for different types of waste'.

'Cradle' and 'grave' are connected by interim steps such as processing, storage, and transport. In IAEA terminology, these steps are subsumed into 'pre-disposal,' the ultimate step being 'disposal,' i.e. the emplacement of the material 'without the intention of retrieval' (IAEA 2018b).

4.2.2 Decisions with impacts on waste arisings

First, the question arises of which types and amounts of waste are to be managed. These types and amounts depend on decisions, both those 'internal' to waste management, and 'external', as evidenced by the following examples:

The arisings of SNF depend on the national nuclear energy programme and its perspectives. The most important decision concerning SNF is whether it is considered a resource to be recycled or, in contrast, a waste to be disposed of ('direct disposal'). If it is considered a resource, there are various options for its management, treatment, and reuse. Amongst these, the only one currently implemented in Western countries is reprocessing using the PUREX method, which separates uranium and plutonium from the fission products. The latter are considered waste, while the former can be used to produce new mixed oxide (MOX) elements or, if this appears to be infeasible, they can later be declared as waste. In any case, reprocessing results in so-called **reprocessing waste**, in particular in HLW containing fission products, but also in other byproducts (e.g. the claddings and structural materials of fuel assemblies, operational wastes). Based on the size and perspectives of the nuclear programme and the existence (or otherwise) of reprocessing capacity, one might distinguish between the following (NEA 2021b, table 2):

- Small programmes, being phased out (e.g. Belgium),
- Small programmes, continuing (e.g. Czech Republic, Hungary),
- Large programmes, no active reprocessing (e.g. Korea, USA),
- Large programmes, active reprocessing (e.g. Japan, France, Russia).

This distinction could further be refined by considering that even programmes which were once considered rather large might be phased out (e.g. Germany), and that programmes with or without active reprocessing might export their SNF for reprocessing abroad or might have done so in the past (e.g. Germany, Japan, Switzerland, Belgium, the Netherlands, Italy; see https://www.orano.group/en/nuclear-expertise/from-exploration-to-recycling/world-leader-in-recycling-used-nuclear-fuels [accessed 27 February 2021]). Usually, contracts related to reprocessing abroad include arrangements for taking back the resulting reprocessing waste, a notable exception being the arrangements between the former Soviet Union and its satellite states: here, the SNF was sent for reprocessing to the Soviet Union, but the reprocessed waste was not returned. When reprocessed waste was taken back, this sometimes applied to essentially all waste types (as in the arrangements between France and Germany), while in other cases the materials containing less radioactivity (LILWs) stayed in the reprocessing country and were 'compensated' for by taking back a slightly higher amount of HLW. Since the policy concerning reprocessing abroad might change over time, some countries, e.g. Germany and Switzerland, have to manage both SNF declared as waste and reprocessed waste. Other countries send all their SNF for reprocessing (e.g. the Netherlands, with the exception of research reactor fuel), or none at all (e.g. Sweden, Canada).

According to the NEA (2021b, p 9), open-cycle, mono-recycle, and multirecycle strategies can be distinguished; open-cycle results in direct SNF disposal, and PUREX reprocessing is an element of mono-recycle strategies. Transmutation can be an element of multirecycle strategies. A thorough description of present and future options for the reuse of SNF and transmutation as well as their implications for radioactive waste management are given in part I, chapter 2 'Nuclear fuel from cradle to grave: existing variants and future options for the fuel cycle and resulting waste types.'

Here, we only name the aspects that are most important when deciding about the reuse (or otherwise) of SNF (Dutton *et al* 2004): obviously, the benefits of reprocessing, or of advanced fuel cycles in general, are strongly dependent on the perspectives of nuclear energy, both the programme in general and the types of reactor in use or envisaged. The security of the fuel supply, and uranium prices on the world market vis-à-vis the costs of reprocessing and fuel production are additional considerations. The exposure of personnel as well as discharge of radionuclides during reprocessing and fuel production need to be taken into account. Furthermore, all the weapons-grade material produced needs to be safeguarded. Compared to direct disposal, the reuse of SNF results in additional transport and storage.

As mentioned above, the reuse of SNF results in specific waste streams that have to be managed, most prominently HLW containing fission products. Currently, the established procedure is to vitrify the liquid waste with the fission products resulting from PUREX reprocessing. Vitrification offers some considerable advantages (see section 4.2.3). On the other hand, complications in the handling of plutonium and MOX result, for example, from instability due to Am-241 build-up and higher decay heat as well as the higher neutron dose of MOX compared to that of UO_2 fuel (McCombie and Isaacs 2010).

The **depleted uranium** remaining after enrichment for the production of nuclear fuel is another material for which a decision has to be made about whether to consider it to be a resource or waste.

Decisions of a very different type, but ones that also influence the inventory of radioactive waste to be managed, might be at stake for materials with a sufficiently low radioactivity content to allow **clearance** (see part I chapter 1 'Radioactivity'). Other materials might be amenable to clearance after being stored for some time ('decay storage') and/or after appropriate processing (e.g. decontamination). No matter whether post-clearance material is to be reused (e.g. scrap metal) or otherwise, in any case, it is no longer subject to radioactive waste management. This is of particular importance when decommissioning nuclear facilities, a process that creates large volumes of residual materials. The question arises which of these materials can be cleared, or cleared after decay storage and/or perhaps extensive treatment processes, such as decontamination, and which materials might be reused. The answers cannot be solely based on economics, but must also account for radiation protection issues (e.g. because treatment might result in workers' exposure, which needs to be justified).

4.2.3 Strategic decisions concerning pre-disposal and disposal

Once it is clear which waste inventory has to be managed, the decision about its 'grave', i.e. its ultimate fate, or the strategic 'end point,' has to be made. A variety of potential end points (from disposal in outer space to emplacement in subduction zones) were considered in studies in the past. It is, however, apparent that for all radioactive materials which are considered waste and the radioactivity content of which does not allow for discharge ('dilute and disperse'), a 'concentrate and contain' approach in dedicated storage and/or disposal facilities is called for. Storage, i.e. the emplacement in a facility with the intent of recovering the materials at some point in time, can be an endpoint for materials with low contents of short-lived radionuclides which allow clearance after a certain timespan (see above). In this case (or in the case of immediate clearance), 'endpoint' means that the material is no longer considered radioactive and therefore does not fall under the rules of radioactive waste management any more. For all other materials, storage is not an endpoint but an interim step. This applies even in cases in which the endpoint is not (yet) known, and the storage time therefore might be undefined. The IAEA advice on the choice of an endpoint is to apply a graded approach in which disposal solutions that differ in terms of the layout, choice, and strength of engineered barriers and the emplacement depth below the Earth's surface are chosen, depending on the hazard potential of the waste (IAEA 2009b, figure 1). For more details about strategy end points, see part I, chapter 7 'Disposal and other conceivable strategy 'endpoints' for different types of waste'.

It is essential for a waste management strategy to foresee a set of disposal facilities which are collectively able to receive the whole waste inventory. Each facility is meant to receive certain types of waste which are characterised using so-called **'waste acceptance criteria' (WAC)**. These criteria, if enforced properly, ensure that only those wastes are delivered to a disposal facility that can be safely handled and disposed of. More generally, WAC can be formulated for all waste management facilities and activities. The issues addressed in WAC might include (IAEA 1983, EC 2018)

- Maximum activity content or concentration (total or nuclide-specific)
- Other radiological issues (e.g. maximum dose rate at the package surface or the absence of contamination)
- Physico-chemical properties (e.g. maximum heat generation, the absence of liquids or gases, the absence of volatile radionuclides, the contents of chemotoxic substances)
- Criticality issues
- Requirements for waste packages (e.g. concerning their handling properties, dimensions, mechanical stability, or corrosion behaviour)
- Requirements concerning declaration, labelling, recording, and tracking

The issues that are relevant for deriving the WAC include technical feasibility, safety during operation and after emplacement in a disposal facility, and considerations about the waste acceptance process (e.g. allowing for controls, inspections, and testing). In particular, facility-specific safety assessments are strongly interrelated with WAC; on the one hand, a facility can only accept waste for which safety can be ensured. For example, radionuclide contents are a central issue when addressing post-operational safety. One could therefore conclude that some WAC are defined based on safety assessments. On the other hand, however, one needs to keep in mind that disposal facilities are not an end unto themselves: if wastes are identified which cannot comply with the WAC, something has to be done about that in terms of either conditioning, reallocation to another facility (if existent), or design changes to the facility resulting in changes in the WAC. Care has to be taken with respect to (over-)conservative approaches in safety assessments with an impact on WAC, since they might result in impractical situations.

In particular, but not only, for HLW and SNF, **container** designs and specifications are of special importance in the context of waste acceptance and strategic choices in general. Many SNF/HLW storage concepts rely strongly on functions that are performed by containers. Often, containers are designed for both storage and transport purposes; for examples including figures see e.g. https://www.gns.de/language=en/23341/casks (for HLW containers) and https://www.gns.de/language=en/24296/mosaik (for LILW) [accessed 20 August 2021]. Multi-purpose containers for storage, transport, and disposal are also under consideration.

All activities prior to disposal ('pre-disposal'), which include **processing**, **storage**, and **transport** (IAEA 2018b), can be seen as means of ensuring that all wastes will end up in a timely manner at the appropriate disposal facilities and meet their WAC. As mentioned above, WAC for pre-disposal facilities (such as storage facilities) or activities (such as transport) might be established and applied.

Processing (i.e. changing physico-chemical characteristics) can further be divided (IAEA 2018b, p 187 f.) into

- **pretreatment** (collection, segregation, chemical adjustment, decontamination),
- **treatment** (volume reduction, activity removal, composition change) and
- **conditioning** (immobilisation, packaging, overpack).

Describing the huge variety of processing approaches and methods, in particular for LILW, its related technologies and waste products, and the associated possible strategic choices would by far exceed the possibilities of this chapter; for basic and overview information see Dutton *et al* (2004) and IAEA (2009a, 2017); selected aspects are addressed in part I, chapters 3 'Other waste origins' and 5 'Waste conditioning and wasteforms'. A description of various LILW processing methods and devices (including figures) is given at https://www.gns.de/language=en/24295/conditioning-methods [accessed 20 August 2021]. Here, we address only a very few selected aspects:

Collection and appropriate documentation and record-keeping is a particular challenge for LILW, especially for small producers. It is necessary to ensure that these wastes are appropriately supervised and that an adequate control, tracking, and delivery system is in place, commonly with one or several storage facilities as destinations, the operation of which might be assigned to waste management organisations or governmental agencies.

Conditioning means producing 'a waste package suitable for handling, transport, storage and/or disposal.' Today, it is generally agreed that only solid or solidified radioactive waste should be disposed of. For the liquids containing fission products that result from reprocessing, vitrification is generally used as the conditioning method. The waste volumes are considerably smaller than the those of SNF that has to be disposed of directly (but of course, the separated uranium and plutonium still exist). The advantages of vitrified waste are that it can be considered to be a standard-quality product, that the borosilicate glass is radiation-resistant and able to contain high concentrations of different radionuclides, and that it is resistant to changes of radionuclide content over time (due to decay). Moreover its chemical durability is high and its solubility is low; both properties are important for the long-term safety of deep geologic repositories (Dutton *et al* 2004, p 54).

For SNF that is to be disposed of directly, the question arises of whether the fuel elements should be disassembled or emplaced in disposal containers as whole assemblies. For these disposal containers, as well as for the ones for vitrified HLW, one can distinguish between self-shielding containers (sometimes called 'lost' containers, since the shielding will be 'lost' in the repository) and containers that need shielding, which can be achieved by appropriate handling and emplacement devices. The choices concerning the conditioning of radioactive waste for disposal are to be made with respect to the operational feasibility and safety of a disposal facility as well as to its long-term safety. Depending on regulations, retrievability considerations might also be of importance.

The strategic choices for **storage** concern the technical layout of the storage facilities, their expected lifetimes, and their locations. Generally, storage is considered to be a temporary measure, i.e. the retrieval of the materials is intended (but the timescale for retrieval might be less clear). A distinction can be made according to the purpose of the storage:

'Storage has long been incorporated as a step in the management of long-lived solid radioactive waste and spent nuclear fuel. Conventionally, storage is seen as an interim step within a management strategy in which the ultimate goal is disposal, or reuse of fissile material and disposal of the residue. Within such a strategy, storage may be needed (a) to allow the levels of radioactivity and heat output to decline before the next step or process of the waste management strategy can be enacted (*decay storage*); (b) to provide stock for an ongoing process, transport step or immediate disposal (*buffer storage*); (c) awaiting a step for which the required facility or transportation capability are not yet available, or awaiting a decision to be made on the next step for a particular waste or material (*interim storage*); (d) for materials that, while not immediately required, have some potential future use or value and, therefore, have not been declared as a waste (*strategic storage*).' (NEA 2006, p 7).

Similarly to the case of processing, the challenges for the storage of LILW are quite different from those for SNF and HLW; the former are mainly related to the wide variety of materials, their origins, or producers, and the necessity to manage them under the appropriate radiation protection regime and surveillance. For the latter, the major concerns are related to shielding, heat removal, criticality safety, the non-proliferation of weapon-grade materials, and security in general. For all waste types and for SNF, containment of the radioactive materials has to be ensured.

The first storage of SNF is decay storage in cooling pools at the reactors. Later, the approaches for SNF storage depend on the plans related to reprocessing or disposal. Technical storage solutions include dry indoor and outdoor storage (with cooling by natural convection, and major safety functions assigned either to the storage containers / casks, or to vault or silo constructions) and storage in pools with cooling by forced water circulation. Storage facilities can be constructed above or under ground. Vitrified HLW from reprocessing is stored in dry facilities, either in buildings for waste containers / casks to which, again, the major safety functions are assigned, or in vault or silo constructions. The lifetime envisaged for such facilities usually lies on the order of decades but, due to the delay in many disposal programmes, longer lifetimes are now often considered and related research and development is being undertaken (NEA 2020b). For more technical details and for research into extended storage times for SNF and HLW see part I, chapter 6 'Storage and transport.'

An important choice for each waste management strategy is that of the locations of the storage facilities. In general, one can distinguish between centralised and local storage. Centralised (national or regional) storage of LILW is linked to the waste collected from small producers. Centralised storage facilities can be planned and constructed independently from, or in connection with, disposal projects—in the latter case, they are particularly intended for buffer storage. Local storage usually means that the materials are stored in the vicinity of the producers, i.e. for SNF, this would be at or close to reactor sites. For waste other than SNF, such local storage—except for short-term buffer storage—only makes sense if the waste arisings are sufficiently high, which might be the case e.g. for certain research facilities.

The choice of local or centralised SNF storage is determined by considerations about 'economics, flexibility, and safety issues associated with transportation' (Dutton *et al* 2004, p 56). In addition, the feasibility of implementing and maintaining safety and security management systems is a consideration when choosing an appropriate approach. In the case of local storage at reactor sites, such management systems, as well as other infrastructure, might be linked to those of the 'hosting' NPPs. The number of NPPs as well as their spatial distribution in a country should be considered as well.

The same factors as those applied to SNF storage are of interest for the storage of reprocessed HLW, which might take place at the same facilities. For countries with their own reprocessing facilities, central HLW storage at these facilities appears to be practical.

In summary, it can be said that the boundary conditions that are most important for SNF/HLW storage approaches are the size and perspective of a country's nuclear programme, the spatial distributions of NPPs, and the approach to reprocessing. The stage of progress of the disposal programme might also be of importance. These boundary conditions can change over time, as can the ownership of the SNF. In particular, an extension of the storage times originally envisaged might be addressed in different ways (e.g. by extending the lifetimes of existing facilities without, or accompanied by, technical changes, or by creating new facilities) and needs to be supported by appropriate management and planning measures, in particular, suitable age management; see NEA (2020b).

Thus, a heterogeneous and complex situation can be observed in the national approaches to HLW/SNF storage. Information about the present situation can be found in (IAEA 1997, 2018a, NEA 2020a) and in table 2.1 (NEA 2020b). Some examples that illustrate this complexity are:

- A prominent example of centralised (national) SNF storage is the Swedish Clab, see https://www.skb.com/our-operations/clab/ [accessed 5 March 2021]. It is also unique in terms of design (underground wet storage).

- In Switzerland, SNF and HLW storage takes place centrally at the ZWILAG facility as well as locally at the Beznau ZWIBEZ facility, see https://www.ensi.ch/en/nuclear-facilities-in-switzerland/waste-management-facilities/ [accessed 5 March 2021].
- In Germany, the HLW/SNF strategy changed from a centralised to a local storage approach (see box 4.3 and in particular, part I, chapter 6 'Storage and transport').
- In the Netherlands, a relatively long-term (100 years and beyond) centralised HLW storage in a vault facility has been implemented, combined with storage of research reactor fuel, depleted uranium, and LILW storage (HABOG), see https://www.covra.nl/en/radioactive-waste/storage/ [accessed 5 March 2021].
- In France, SNF is being stored at the NPP sites before being sent to the ORANO reprocessing plant in La Hague (with wet SNF storage prior to reprocessing). The reprocessed waste (HLW and other) is centrally stored there while awaiting disposal, see https://www.iaea.org/sites/default/files/6rm-france.pdf [accessed 5 March 2021].
- Strategic changes occurred or might occur for various reasons, including delays in national disposal programmes (Germany, Italy, Spain, Switzerland, UK, USA), changes related to reprocessing (potentially in Hungary), and new reactor builds (Finland, Poland).

Transport can be seen as just a means for connecting the strategic elements mentioned above. There are, however, several issues which require specific attention: the transports of radioactive materials are subject to a multitude of regulations that assign responsibilities to various regulators. In the case of transboundary transports, the situation becomes even more complex. Specific requirements exist, in particular, with respect to container design and declaration issues. Safety and security concerns are different from those for stationary facilities. Last but not least, such transports might attract specific attention in political conflicts about nuclear issues and, as in the case of Germany, become subject to protests and blockade action.

Bibliography

BMU (Bundesministerium für Umwelt, Naturschutz und nukleare Sicherheit—Federal Ministry for Environment, Nature Conservation and Nuclear Safety) 2015 *National Programme. Programme for the Responsible and Safe Management of Spent Fuel and Radioactive Waste* www.bmuv.de/en/download/national-programme

CoRWM (Committee on Radioactive Waste Management) 2006 *Managing our Radioactive Waste Safely. CoRWM's Recommendations to Government* https://gov.uk/government/uploads/system/uploads/attachment_data/file/294118/700_-_CoRWM_July_2006_Recommendations_to_Government_pdf.pdf

Dutton M *et al* 2004 *The Comparison of Alternative Waste Management Strategies for Long-Lived Radioactive Wastes (COMPAS Project)* EUR 21021 EN, CEC, Luxembourg https://op.europa.eu/en/publication-detail/-/publication/c5f948f4-7c05-4f94-8bdf-66342864d973

EC (European Commission) 2018 *Waste Acceptance Criteria and Requirements in Terms of Characterisation* Theramin Thermal treatment for Radioactive Waste Minimisation and

Hazard Reduction, deliverable D4.1 https://igdtp.eu/wp-content/uploads/2019/07/THERAMIN_D4_1_Waste_Acceptance_Criteria_and_requirements.pdf

Endlagerkommission (Kommission Lagerung hoch radioaktiver Abfallstoffe) 2016 *Abschlussbericht: Verantwortung für die Zukunft. Ein faires und transparentes Verfahren für die Auswahl eines nationalen Endlagerstandortes* (in German—Commission on the Storage of High-Level Radioactive Materials *Final Report. Responsibility for the Future. A Fair, Transparent Procedure for the Selection of a National Disposal Site*) https://bundestag.de/endlager-archiv/blob/434430/bb37b21b8e1e7e049ace5db6b2f949b2/drs_268-data.pdf English translation under https://bmu.de/fileadmin/Daten_BMU/Download_PDF/Endlagerprojekte/bericht_kommissionn_lagerung_radioaktiver_abfallstoffe_bf.pdf

EU (European Union) 2011 Council Directive 2011/70/EURATOM of 19 July 2011 Establishing a Community Framework for the Responsible and Safe Management of Spent Fuel and Radioactive Waste *Off. J. EU* L 199/48 http://data.europa.eu/eli/dir/2011/70/oj

IAEA (International Atomic Energy Agency) 1983 *Characteristics of Radioactive Waste Forms Conditioned for Storage and Disposal: Guidance for the Development of Waste Acceptance Criteria* IAEA-TECDOC-285 www-pub.iaea.org/MTCD/Publications/PDF/te_285_web.pdf

IAEA (International Atomic Energy Agency) 1997 *Joint Convention on the Safety of Spent Fuel Management and on the Safety of Radioactive Waste Management* https://iaea.org/sites/default/files/infcirc546.pdf

IAEA (International Atomic Energy Agency) 2009a *Policies and Strategies for Radioactive Waste Management* IAEA Nuclear Energy Series No. NW-G-1.1 https://www-pub.iaea.org/MTCD/Publications/PDF/Pub1396_web.pdf

IAEA (International Atomic Energy Agency) 2009b *Classification of Radioactive Waste* General Safety Guide IAEA Safety Standards Series No. GSG-1 https://www-pub.iaea.org/MTCD/Publications/PDF/Pub1419_web.pdf

IAEA (International Atomic Energy Agency) 2017 *Selection of Technical Solutions for the Management of Radioactive Waste* IAEA-TECDOC-1817 https://www-pub.iaea.org/MTCD/Publications/PDF/TE-1817_web.pdf

IAEA (International Atomic Energy Agency) 2018a *Sixth Review Meeting of the Contracting Parties to the Joint Convention on the Safety of Spent Fuel Management and on the Safety of Radioactive Waste Management (Joint Convention)* https://iaea.org/events/sixth-review-meeting-of-the-contracting-parties-to-the-joint-convention-on-the-safety-of-spent-fuel-management-and-on-the-safety-of-radioactive-waste-management-joint-convention

IAEA (International Atomic Energy Agency) 2018b *IAEA Safety Glossary. Terminology Used in Nuclear Safety and Radiation Protection. 2018 edn* https://www-pub.iaea.org/MTCD/Publications/PDF/PUB1830_web.pdf

McCombie C and Isaacs T 2010 The Key Role of the Back-End in the Nuclear Fuel Cycle *Multinational Approaches to the Nuclear Fuel Cycle* ed C McCombie, T Isaacs, N Bin Muslim, T Rauf, A Suzuki, F von Hippel, E Tauscher (American Academy of Arts and Sciences) https://amacad.org/publication/multinational-approaches-nuclear-fuel-cycle

NEA (Nuclear Energy Agency) 2006 *The Roles of Storage in the Management of Long-lived Radioactive Waste. Practices and Potentialities in OECD Countries* OECD, NEA No. 6043 https://oecd-nea.org/jcms/pl_14068/the-roles-of-storage-in-the-management-of-long-lived-radioactive-waste

NEA (Nuclear Energy Agency) 2020a *Radioactive Waste Management and Decommissioning Programmes in NEA Member Countries. Country Profiles and Reports* https://oecd-nea.org/jcms/pl_33688/radioactive-waste-management-programmes-in-nea-member-countries

NEA (Nuclear Energy Agency) 2020b *Storage of Radioactive Waste and Spent Fuel* OECD, NEA No. 7406 https://oecd-nea.org/jcms/pl_38031/storage-of-radioactive-waste-and-spent-fuel

NEA (Nuclear Energy Agency) 2021a *Ensuring the Adequacy of Funding for Decommissioning and Radioactive Waste Management* OECD, NEA No. 7549 https://www.oecd-nea.org/jcms/pl_59705/ensuring-the-adequacy-of-funding-for-decommissioning-and-radioactive-waste-management

NEA (Nuclear Energy Agency) 2021b *Strategies and Considerations for the Back End of the Fuel Cycle* OECD, NEA No. 7469 https://oecd-nea.org/jcms/pl_55928/strategies-and-considerations-for-the-back-end-of-the-fuel-cycle

NRCAN (Natural Resources Canada) 2015 *Radioactive Waste Management Programmes in OECD/NEA Member Countries: Canada* https://oecd-nea.org/jcms/pl_33705/canada-profile-web

OPG (Ontario Power Generation) 2021 *Our Projects. Deep Geologic Repository* https://opg.com/strengthening-the-economy/our-projects/the-deep-geologic-repository/

Röhlig K-J, Chaudry S and Plischke E 2019 *Entsorgungsoptionen für radioaktive Reststoffe: Interdisziplinäre Analysen und Entwicklung von Bewertungsgrundlagen (ENTRIA, 2013–2018): Abschlussbericht. Ergebnisse und Leistungsbilanz (in German, Disposal Options for Radioactive Residues: Interdisciplinary Analyses and Development of Evaluation Principles (ENTRIA, 2013–2018): Final Report. Results and Achievements' Summary)*

SKB (Svensk Kärnbränslehantering AB, Swedish Nuclear Fuel and Waste Management Company) 2021 *A Safe Final Repository System* https://skb.com/our-operations/the-swedish-system/

Streffer C, Gethmann C F, Kamp G, Kröger W, Rehbinder E, Renn O and Röhlig K-J 2011 *Radioactive Waste. Technical and Normative Aspects of its Disposal* (Berlin: Springer)

IOP Publishing

Nuclear Waste
Management, disposal and governance
Klaus-Jürgen Röhlig

Chapter 5

Waste conditioning and wasteforms

Michael I Ojovan

5.1 Conditioning

Conditioning produces a waste package suitable for handling, transport, storage and/or disposal and may include the conversion of liquid waste to a solid wasteform, the additional immobilization of some solid waste, and the packaging of the wasteform into containers and overpacks if necessary. 'Wasteform' means the solid waste product in its physical and chemical form after treatment and/or immobilization prior to packaging. The constituents of the waste can be immobilized into a wasteform material through two processes, either via chemical incorporation that binds them into the material at the atomic scale, or via encapsulation, in which they are physically surrounded and isolated by the material. The immobilization of radioactive waste via solidification, embedding, or encapsulation that aims to obtain a stable wasteform minimizes the potential for the migration or dispersion of radionuclides into the environment during storage, handling, transport, and disposal. Within the repository, the wasteform is one part of a multiple engineered barrier system. During storage and transportation, the wasteform is the primary barrier preventing the release of radionuclides into the environment, while during post-closure disposal, the wasteform will reduce the release of radionuclides from breached and compromised containers that could result from disruptive phenomena.

5.2 Wasteforms

Choosing a suitable wasteform for nuclear waste immobilization is not straightforward, and durability is not the sole acceptance criterion, as priority is given to reliable, simple, rugged technologies and equipment, which may have advantages over complex or sensitive equipment and processes. A variety of matrix materials and techniques are available for immobilisation. The choice of immobilization technology depends on the physical and chemical nature of the waste and the acceptance criteria for the storage and disposal facility to which the waste will be consigned. Several factors should be considered when selecting a wasteform material

doi:10.1088/978-0-7503-3095-4ch5

to immobilize a specific waste stream. The key considerations include the following (IAEA 2017, NRC 2011, Ojovan *et al* 2019):

- *Waste loading*: The wasteform should be able to accommodate a significant amount of waste to minimize volume, thereby minimizing the space needed for storage, transportation, and disposal.
- *Ease of production*: Fabrication of the wasteform should be accomplished under reasonable conditions, including low temperatures and, ideally, in an air atmosphere, using well-established methods to minimize worker dose and the capital cost of plant.
- *Durability:* The wasteform should have a low rate of dissolution when in contact with water to minimize the release of radioactive and chemical constituents.
- *Radiation stability*: The wasteform should have a high tolerance to radiation effects from the decay of radioactive constituents. Depending on the types of constituent being immobilized, the wasteform could be subjected to a range of radiation effects, including ballistic effects from alpha decay and ionizing effects from the decay of fission product elements.
- *Chemical flexibility:* The wasteform should be able to accommodate a mixture of radioactive and chemical constituents with a minimum formation of secondary phases that can compromise its durability.
- *Natural analogues:* Since direct laboratory testing of the wasteforms over the relevant time scales for disposal (typically 10^3–10^6 years) is not possible, the availability of natural mineral or glass analogues may provide important clues about the long-term performance of the material in the natural environment, thereby building confidence in the extrapolated behaviour of the wasteform after disposal.
- *Compatibility with environment:* The wasteform should be compatible with the near-field environment of the disposal facility. The near-field environment provides the physical and chemical conditions that are favourable for maintaining wasteform integrity over extended periods, which helps to slow the release of the constituents and their transport out of the facility.

A host of regulatory, process, and product requirements has led to the investigation and adoption of a variety of matrices and technologies for waste immobilization. The resistance of the wasteform to aqueous corrosion and the release of radionuclides in the disposal environment—chemical durability—is a critical parameter. The main immobilization technologies currently available commercially which have been demonstrated to be viable on an industrial scale are cementation and vitrification, while bitumen and polymeric material immobilisation has been used to a smaller extent (Ojovan *et al* 2019). Ceramification is a prospective technology that will most probably be deployed in addition to vitrification, as it provides a higher degree of radionuclide confinement, which is especially important for long-lived actinides. Table 5.1 shows the generic features and limitations of the main wasteform materials currently used on an industrial scale.

Table 5.1. Features and limitations of the main wasteforms. Reproduced by permission of the IAEA from IAEA (2017).

Wasteform	Features	Limitations	Secondary waste
Glasses	A proven method for conditioning liquid high-level waste (HLW) as well as intermediate-level waste (ILW) and low-level waste (LLW). High flexibility in terms of the glass formulation range. High reliability of the immobilization process. High glass throughput. High durability of the final wasteform. Small volume of the resulting wasteform.	High initial investment and operational costs. Complex technology, requiring highly qualified personnel. Need to control off-gases. High specific energy consumption.	Off-gases. Filters. Scrub solutions. Used melters.
Ceramics	It is possible to incorporate higher levels of actinides than for borosilicate glass. Wasteform can be more durable than glass. Expected to be suitable for long-term isolation, since it simulates natural rocks.	Limited experience. Most efforts have been research-based. The ceramic has to be tailored to the nuclear waste's composition.	Filters. Off-gases. Scrub solutions.
Glass-composite materials	Combine features of both crystalline and glassy materials. Higher waste loading. Higher compatibility. Higher stability compared to glasses.	Limited experience.	Off-gases. Filters. Scrub solutions. Used melters.
Cements	Widely-used method for a variety of LLW and ILW. High flexibility. Low cost. Simplicity of process. Low temperature precludes volatile emissions. High radiation stability, impact, and fire resistance of wasteforms.	Increase in volume (low waste loading). Low retention of some fission and activation products. Poor compatibility with organic materials and high-salt content solutions.	None.
Bitumen	Mostly used for LILW, chemical precipitates, low heat, and low alpha wastes. High flexibility.	Sensitivity to some components. Low fire resistance.	Filters.

(*Continued*)

Table 5.1. (*Continued*)

Wasteform	Features	Limitations	Secondary waste
	High compatibility with organic materials. High waste loading. Lower leaching rate compared with cements.		
Metals	Extensively proven technology for conditioning of metallic waste. The product is typically homogeneous and stable.	Pre-sorting is usually required due to dedicated melt furnaces and differences in melt temperatures of different metals.	Off-gases. Slag.

5.3 Wasteform (matrix) materials

Vitreous wasteforms (glasses) are being used worldwide to immobilise high-level nuclear waste (HLW) resulting from the reprocessing of spent nuclear fuel (SNF) from operating reactors and deployed for legacy and operational radioactive wastes from the nuclear fuel cycle (Ojovan 2011, Donald 2010). The immobilisation process known as vitrification is a continuous process capable of handling high-volume waste streams. Vitrification is currently deployed on a large scale for lower-activity waste streams and has become the reference process for the conditioning of HLW. Given the positive results obtained for the vitrification of HLW, several projects are under way for the vitrification of slurries, low- and intermediate-level solid waste, and mixed waste. Vitrification offers volume reduction, the destruction of organic constituents including hazardous materials, the immobilization of radioactive and hazardous components, and advantages for storage, transportation, and disposal. Vitrification processes are sufficiently robust, which means that they accept almost any waste after a minimum of up-front characterization, with reproducible characteristics of the end vitreous product and acceptable off-gases during immobilisation. Vitrification can also be performed *in situ* as a special case e.g. with legacy waste or contaminated soil.

Crystalline ceramics are inorganic, non-metallic solids that contain one or more crystalline phases (Ojovan 2011, Donald 2010). Single-phase crystalline ceramics can be used to immobilize separated radionuclides (e.g. Pu-239) or more chemically complex waste streams (e.g. HLW). In the latter case, the atomic structure of the ceramic phase should have multiple cation and anion sites that can accommodate the variety of radionuclides present in the waste stream. These materials are potentially attractive for immobilizing long-lived alpha-emitting actinides, such as plutonium, neptunium, and americium; however, some of these materials are susceptible to radiation damage effects associated with alpha decay from actinides. Multiphase crystalline ceramics (e.g. *Synroc*) consist of an assemblage of crystalline phases. Individual phases are selected for the incorporation of specific radionuclides, and the proportions of the phases vary, depending on the composition of the waste stream. An individual phase can host one or more radionuclides, including solid solutions of radionuclides. However, not all phases will host radionuclides. Ceramic materials and fixation methods of fixation have been developed or are being developed. Ceramic products are crystalline in nature and therefore thermodynamically stable, though they are more sensitive to radiation damage.

Glass-composite materials (GCM's) contain both crystalline and glass phases (Ojovan *et al* 2019, Ojovan 2011, Donald 2010). Depending on the intended application, the major component may be a crystalline phase with a vitreous phase acting as a bonding agent. Alternatively, the vitreous phase may be the major component, with the particles of a crystalline phase dispersed in the vitreous matrix. GCMs can be formed by a number of processes, including melt crystallization (controlled or uncontrolled), multiple heat treatments, or by the encapsulation of ceramic material in glass. GCMs offer several potential advantages over glass for use as wasteform materials, including increased waste loadings, increased wasteform density and smaller disposal volumes. These wasteforms can also be used to immobilize glass-immiscible components, such as sulphates, chlorides, molybdates, and refractory materials that have very high melting temperatures. They can also be used to immobilize long-lived radionuclides such as actinides by incorporating them into the more durable crystalline phases; short-lived radionuclides, such as most fission products can be accommodated in the less durable vitreous phase.

Cements are inorganic materials that set and harden as a result of hydration reactions (Ojovan *et al* 2019, Abdel Rahman *et al* 2015). Cements are used to immobilize waste that has relatively low levels of radioactivity i.e. mainly low- and intermediate-level radioactive wastes (LLW and ILW). Higher-activity wastes can result in radiolysis and the production of hydrogen gas from the breakdown of water or hydroxyl groups in the cement. Cementation is low cost and forms a major part of both solid and liquid (mainly aqueous) LLW and ILW immobilization technologies. The range of applicability of cements should be considered in view of the characteristics of the environment and of the initial waste. The cement may display pH buffering properties and consequently control the mobility of most of the radionuclides in the disposal environment. The quality of cemented wasteforms is continuously improving. They are also efficient for the immobilization of alpha-bearing waste. Their disadvantages relate to their relatively high porosity and leachability for some radionuclides of the product as well as the volume increase rather than decrease that may result.

Bitumen is a viscous hydrocarbon and a major component of asphalt which has been used to solidify and stabilize radioactive materials (Ojovan *et al* 2019). Bitumen immobilizes waste by encapsulation; it does not bind the waste chemically. The advantages of bitumen as a wasteform are simplicity of production, low operating cost, and leach-resistant characteristics. However, bitumen does not perform well with dehydrated salts, such as sodium sulphate, sodium nitrate, magnesium chloride, and aluminium. It can also be a fire hazard, especially when oxidizing wastes such as nitrates are involved.

In so-called once-through nuclear fuel cycles (NFC), the spent nuclear fuel is considered waste (SFW) when it is not reprocessed. Its conditioning consists of volume optimization via rearrangement of the fuel rods and envelopment in a multi-component barrier consisting of various *metals*, such as copper and the packaging canister (Ojovan 2011). Several different types of metallic material, including lead, have been studied as potential wasteforms. Like crystalline ceramics, metal wasteforms can consist of single- or multiple-phase assemblages, and the wasteform itself can be granular or monolithic. Metallic wasteforms can be fabricated using sintering or casting. Each of these techniques has drawbacks; in particular, it can be difficult to find metal compositions and processes that effectively wet and encapsulate dispersed phases or fines.

5.4 Waste packages

The waste package is the product that includes the wasteform and the container plus the internal barriers such as the absorbing materials and liners used. Their radiological and physicochemical properties, roles, and significance in the waste management process are illustrated in table 5.2. The waste container is designed to contain, physically protect, and/or radiologically shield the wasteform during the various activities involved during the period from conditioning until emplacement and the closure of a disposal facility. In some cases, the container also plays a role in the near-field containment of the radionuclides for a certain period after the closure of the disposal facility. Waste packages evolve during storage, and those changes that affect the safety function need to be understood and controlled to satisfy the regulators of the safety of the store and waste (Lee *et al* 2013).

Table 5.2. Wasteform and waste package properties. Reproduced by permission of the IAEA from IAEA (2017).

Property		Role and significance of the property
Radiological properties	– Total activity – Radionuclide composition – Surface dose rate – Surface contamination – Thermal power – Radiation stability – Fissile content	– To comply with licensing limits of a disposal facility – To meet transport regulations Optimize final wasteform activity determination methods. – Dimension, density, shape, etc. should support an efficient determination process: ○ Define handling conditions. ○ Select appropriate conditioning and packaging technologies. ○ Minimize waste generation (including secondary waste). ○ Design the facility for either long-term storage or disposal. • Optimizing radiological protection, design of shielding, etc. ○ To prevent inadvertent criticality

Physical and chemical properties	– Mass and weight – Structural and dimensional stability – Permeability and porosity – Density – Presence of voids – Mechanical Strength/ Load Resistance – Impact resistance – Homogeneity – Chemical stability (leachability) – Chemical composition – Corrosivity – Explosiveness – Gas generation – Toxicity – Fire resistance	– To maintain the physical integrity of the wasteform and/or package, during handling, transportation, storage, and disposal – To prevent deformation or cracking of the wasteforms and packages A categorization of conditioned waste based on the physical and chemical properties is an adequate tool to define: ○ Waste-conditioning techniques, technologies, and equipment design. ○ Operational processing parameters. ○ Wasteform and package compatibility. Waste packages. ○ Dimensions, weight, materials, etc. ○ Handling and transport conditions. ○ Compatibility between the wasteform and the WAC. ○ Chemical, thermal, structural, mechanical, and radiation stability. – To prevent the dispersion of long-lived radionuclides in the repository – To undertake special measures to preclude ignition during transportation and storage

Bibliography

Abdel Rahman R O, Rakhimov R Z, Rakhimova N R and Ojovan M I 2015 *Cementitious Materials for Nuclear Waste Immobilisation* (Chichester: Wiley)

Donald I W 2010 *Waste Immobilisation in Glass and Ceramic Based Hosts* (Chichester: Wiley)

IAEA (International Atomic Energy Agency) 2017 *Selection of Technical Solutions for the Management of Radioactive Waste* IAEA TECDOC-1817

Lee W E, Ojovan M I and Jantzen C M 2013 *Radioactive Waste Management and Contaminated Site Clean-Up: Processes, Technologies and International Experience* (Cambridge: Woodhead)

NRC (National Research Council) 2011 *Waste Forms Technology and Performance: Final Report* (Washington, DC: National Academies Press)

Ojovan M I 2011 *Handbook of Advanced Radioactive Waste Conditioning Technologies* (Cambridge: Woodhead) pp 67–135

Ojovan M I, Lee W E and Kalmykov S N 2019 *An Introduction to Nuclear Waste Immobilisation* 3rd edn (Amsterdam: Elsevier)

Chapter 6

Storage and transport

Maik Stuke, Mathias Häring and Jörn Becker

6.1 Storage of radioactive waste

The life cycle of generated radioactive waste follows a strategy with well-defined steps (see part I, chapter 4, 'Waste management policy and strategy'). Prior to its final destination, radioactive waste is generally stored in interim storage. The interim storage of radioactive waste is defined as a temporary process to ensure that maintenance takes place and to allow the further following steps of the waste management strategy to take place (OECD 2006). This involves retrievability, confinement, isolation, environmental protection, and monitoring during the entire storage period (IAEA 2006). Generally, all storage systems consist of two major components: the waste package and the storage facility. The waste package includes the form taken by the waste and the enclosing container. The storage facility includes the entire storage structure needed to achieve safe operation, e.g. the infrastructure required for handling, monitoring, ventilation, and cooling.

The end point of any nuclear waste management strategy is the disposal of the waste. This is the subject of part I, chapters 4 and 7.

Interim storage facilities are generally designed to ensure the safe storage of low- and intermediate-level waste (LILW) and high-level waste (HLW) pending the availability of final disposal. Specially designed interim storage facilities at, near, or below ground level are currently operated in many countries. LILW is most commonly stored in suitable canisters above the ground. In addition to high-level vitrified waste, the vast majority of HLW consists of used or spent nuclear fuel (SNF) from commercial nuclear power reactors. Since the form of the waste defines the storage package and facility, it is worth discussing some of the differences between used nuclear fuels from different reactor types and operational histories. This becomes especially important when we consider the material degradation processes that occur during the long-term storage of spent nuclear fuel, which is an emerging reality due to the delays involved in final disposal (see part I, chapter 8). The following provides an overview of the storage of used nuclear fuel in wet and

doi:10.1088/978-0-7503-3095-4ch6

dry storage facilities, and discusses some implications of the long-term storage of used nuclear fuel.

6.1.1 Used nuclear fuel and its storage

Currently, more than 440 commercial power reactors are operating worldwide, producing around 10 000 metric tons of heavy metal in form of used nuclear fuel each year. The vast majority of reactors are light water reactors (LWRs) or heavy-water reactors (HWRs) that use uranium oxide fuels. The majority of LWR fuel assemblies are usually enriched in the fissile ^{235}U to 3 to 5 wt.%[1] and the averaged fuel assembly burnup is in the range of 30 to 65 GWd / tHM.[2] Generally, HWR fuel contains less ^{235}U and has smaller burnups. An increase in burnup is usually correlated with increasing enrichment.

If a country reprocesses used nuclear fuel in a closed or partially closed fuel cycle, LWR burnups are typically in the range of 30 to 40 GWd / tHM. However, only a few countries currently follow a (partially) closed-cycle strategy, e.g. France, Russia, and China. Most countries operating nuclear power plants do not, or no longer, reprocess used nuclear fuel. Following the once-through strategy allows more of the energy in a fuel assembly to be used, since the composition of the isotopes in the fuel does not have to be in the favorable regime for reprocessing, i.e. around 35 GWd / tHM. Thus, higher burnups and enrichments are used.

During the irradiation of UO_2 fuel in the reactor, the isotopic content of the fuel changes. Neutron capture leads to a build-up of plutonium, which, depending on the isotope, plays a further part in the fission and energy generation process. A fuel mixture of UO_2 and PuO_2 can also be achieved by blending PuO_2 from reprocessing and UO_2 powder to form so-called mixed-oxide (MOX) fuel. LWR MOX fuels typically contain less than 10 wt.% Pu. The isotopic content of a nuclear fuel after irradiation determines its characteristics e.g. decay heat, reactivity, and toxicity and thus directly impacts the storage requirements.

Prior to its service in the reactor, the nuclear fuel is commonly called *fresh fuel*. For the production of the fuel assemblies, the nuclear fuel is sintered in ceramic pellets and stacked in metallic tubes called *fuel rods*. The latter are assembled into fuel assemblies using suitable spacer grids and top and bottom nozzles. The dimensions and geometric arrangements of the rods in the assembly depend on the reactor design. The majority of reactors are LWRs that use H_2O as the moderator for neutrons and for core cooling and heat transport. Due to the neutron capture cross section of hydrogen, LWR fuel has to be enriched in ^{235}U to enable a sufficient chain reaction to take place. LWRs include three major reactor concepts, namely pressurized light water reactors (PWR), the Russian water cooled and moderated reactor (VVER), and the boiling water reactor (BWR). PWRs operate in

[1] Wt.% stands for *percentage by weight*.

[2] The burnup of a fuel assembly describes how much energy is extracted from the nuclear fuel, commonly expressed in gigawatt days per (metric) ton of heavy metal (GWd / tHM).

a pressure and temperature regime that prevents water from boiling. The majority of LWR reactors are PWRs that use a multitude of different fuel assembly designs.

Boiling water reactors generally operate at lower pressures, which allow the cooling water to boil. Due to the resulting density variation of the water, the neutron moderation varies. To achieve an improvement in the uranium utilization, axially distributed enrichments and the use of so-called burnable poisons such as gadolinium can be used in the design of fuel assemblies. As with PWRs, a variety of different BWR fuel types and assemblies exist.

Both PWR and BWR fuel rods are typically around 4 m high and arranged in square lattices of 8×8 to 10×10 rods for BWR assemblies and 14×14 to 18×18 for PWR assemblies. VVER assemblies are typically arranged in a hexagonal lattice. The fuel assemblies for different reactor types are commonly stored separately.

Further reactor types that use different fuel concepts are the pressurized heavy-water (D_2O) reactors, e.g. the CANada Deuterium Uranium (CANDU) reactor, advanced gas-cooled reactors (AGR), or the graphite-moderated RBMK reactors. The following will focus on the storage of the used fuel from LWRs. For an overview of some reactor designs, see IAEA (1996).

The safe storage of used nuclear fuel is achieved by maintaining at least two independent barriers between the fuel and the environment. For fresh fuel, the radioactive particles are bound in the pellet. However, after the irradiation in the reactor, the chemical and physical pellet structure changes and radioactive particles are released inside the fuel rod. Thus, the surrounding cladding is considered to be the primary barrier for an undamaged fuel rod. The secondary barrier is the entire confinement structure of the storage system, e.g. the cask. The confinement system is designed to withstand damaging events e.g. extreme natural events, such as earthquakes and terrorist attacks.

The basic safety objectives during the storage of used nuclear fuel are defined in the IAEA Safety Standard Series (IAEA 2002). The following four criteria could be seen as basic (Vitkova *et al* 2006):

- Radiation protection.

  The primary requirement of a storage system is the separation of the radioactive waste from the biosphere and the protection of the environment from radiation exposure. The system has to effectively shield the gamma and neutron radiation. Furthermore, the systems have to be sufficiently sealed to retain any radioactive particles. All systems are designed following the ALARA (As Low As Reasonably Achievable) principles for radiation exposure.
- Maintaining subcriticality.

  The effective neutron multiplication factor k_{eff} has to remain below a defined value to avoid any uncontrolled neutron chain reaction. Most regulations require $k_{eff} < 0.95$ on the assumption that the fuel is unirradiated (fresh fuel assumption). The subcriticality has to be ensured not only by the design features of the storage system but also by the handling systems used under normal conditions.

- Removal of residual heat.

  The effective removal of decay heat from the storage system can be achieved either passively (through natural convection or thermal radiation) or actively (by forced convection). Generally, temperature limits are defined in regulations to avoid the thermal degradation of materials and to maintain the effectiveness of the storage components. An example of these limits is the definition of a maximum peak cladding temperature during storage and handling, which is limited, for example, to 370 °C in Germany and 400 °C in the United States of America.
- Retrievability.

  After interim storage, each individual assembly or encased assembly needs to be retrievable.

Storage concepts that fulfil these criteria can be divided into two categories: wet and dry storage, which are further described in the following sections.

6.1.2 Wet storage

After the fuel assemblies have been irradiated in the reactor, the spent fuel still contains a large amount of fissile material and radioactive isotopes and the ongoing radioactive decay still generates high power and radiation. For safe fuel handling, all LWR reactor designs include a fuel pool or pond to store the assemblies until the decay heat has decreased sufficiently to allow further transport or dry storage to take place.

After a reactor shutdown, the reactor vessel is opened. For an efficient use of the nuclear fuel, some of the fuel assemblies in the core are replaced by fresh fuel. A refuelling machine transfers the used fuel assemblies under a sufficient shielding height of the waterline from the reactor core into the fuel pool and loads the fresh fuel assemblies into the reactor. Generally, the wet storage always has to guarantee enough free capacity to enable a complete unloading of the core if necessary. Depending on the PWR reactor design, the fuel pools are either inside the reactor's primary containment, as is the case, for example, in the KONVOI, EPR, and VVER-1000 designs, or adjacent to it in a separate building. For BWRs, the fuel pool is generally outside the primary containment of the reactor (IAEA 1996).

Spent fuel pools are generally large and robust steel-lined concrete structures with walls thicker than one meter. The approximately four-meter high fuel assemblies are stored in metallic racks made of stainless steel, which often include additional neutron absorbers. Typically, seven meters of water are provided above the assemblies to ensure that effective shielding from radiation is provided, even during the transfer operation of a fuel assembly above the stored fuel assemblies[3].

[3] An increased height of water also increases the thermal mass of the cooling water. This leads to additional safety margins if the temperature increases, e.g. due to a failed assembly or a loss-of-cooling accident.

Generally, the water inside the pool is demineralized for BWR reactor designs and borated for PWRs. Some typical chemical values are given e.g. in IAEA (2002) as a pH of 4 to 6, chloride, sulphate, and fluoride concentrations each lower than 0.1 ppm, and Na^+ and Ca^{2+} each less than 0.5 ppm. Close monitoring of the water chemistry is necessary during the operation of wet storage.

The design details of the fuel pools may vary not only between different reactor technologies but also between individual design series. For an overview of selected spent fuel pool designs and storage rack types, see appendix A of OECD (2015).

If the design of the fuel pool allows the storage of the entire inventory of used fuel generated during the reactor's lifetime, the fuel can remain in the pool until a final repository is operational. However, most reactor designs include only limited space, assuming that fuel is removed for the purpose of reprocessing or further dry storage. In these cases, the fuel is stored in the pool for a period from several months to decades until the majority of decay power and radioactivity (compared to the time of unloading) has decayed. The storage duration is linked to the next step in the waste management strategy and to the isotopic content of the fuel. For example, due to their slower decay heat decrease, MOX fuel assemblies need to remain longer in the pool to cool compared to UO_2 fuel. Another factor is the burnup of the fuel assembly; the higher the burnup, the longer the assemblies remain thermally hot (see figure 6.1).

If the capacity of the fuel pool reaches its limit, the fuel needs to be stored in so-called away-from-reactor (AFR) storage facilities. An example of further AFR wet storage is the central interim storage facility for spent nuclear fuel, Clab, in Sweden (SKB 2021). The entire used nuclear fuel of Sweden (up to 8000 metric tons) can be stored in eight fuel pools located in two rock caverns approximately 40 meters below the surface on a peninsula.

6.1.3 Dry storage

In comparison to wet storage, dry storage uses an inert gas, e.g. helium or nitrogen, to further cool the fuel assemblies and concrete or metallic structures to shield the radiation. As mentioned above, the capacity of a fuel pool might reach its operational limit. As an alternative to further wet storage in AFR fuel pools, the used fuel can be stored in dry casks while awaiting further transport to a reprocessing facility or a final repository.

The dry storage casks are loaded underwater inside the fuel pool to ensure safe handling and to use the water as shielding (see figure 6.2). Once a cask is fully loaded, all the water inside the cask has to be pumped out and the residual water has to be dried. A variety of different drying processes exist, e.g. vacuum drying with and without the application of external heating to the cask. Generally, reducing the pressure inside the cask via a vacuum pump leads to an increase of the fuel rod cladding temperature and evaporation of the residual water. Due to the low heat transfer during the vacuum process, the cask is backfilled in cycles with helium gas to limit the cladding temperature and thus to suppress degradation mechanisms such as creep, which might lead to cladding failure and the potential loss of the first safety barrier. Once the cask has been

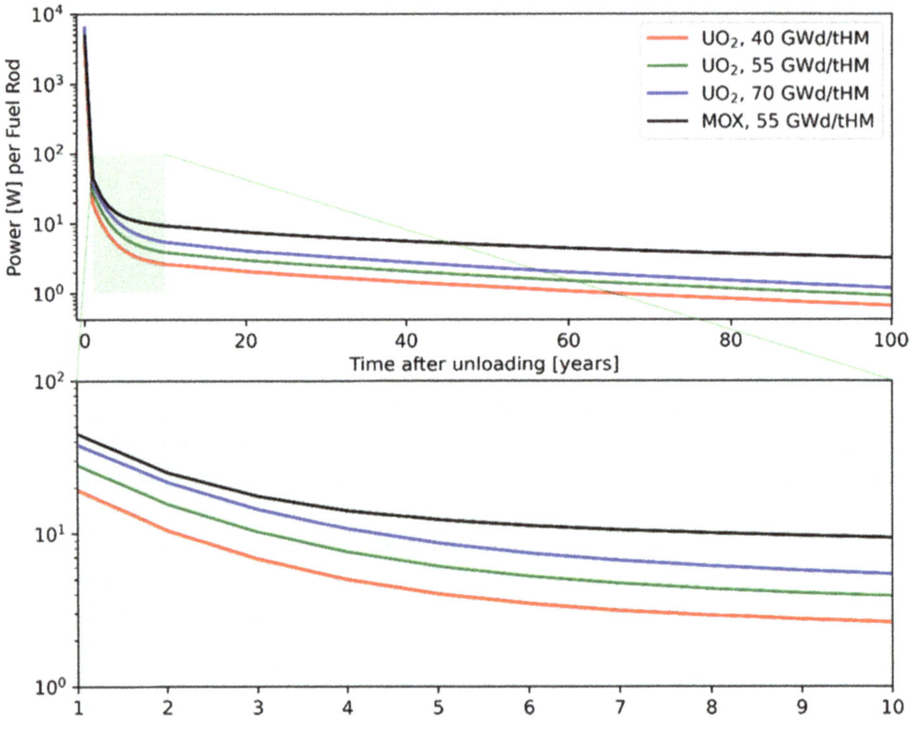

Figure 6.1. The decay power in watts of the hottest fuel rod of a generic 18x18–24 Konvoi-type PWR fuel assembly with UO_2 (4.65% enrichment, different burnups) and MOX (3.3% ^{239}Pu and ^{241}Pu) fuels, assuming a typical burnup power profile. The upper picture shows the evolution of UO_2 and MOX fuels for 100 years after unloading the fuel from the reactor while the lower picture is an enlargement of years one to ten.

sufficiently dried and filled with the inert gas, it is closed and can be transported to the dry storage facility. Most dry storage facilities are on the reactor site to avoid long-distance transportation.

Dry storage systems can be divided into three categories: dry storage in vaults, in silos, and in specially designed casks. The latter can be further divided into single, dual, and multi-purpose casks.

6.1.3.1 Vaults and silos

The storage of used nuclear fuel in vaults uses large concrete buildings to shield the radiation. Inside the concrete building, a number of cavities exist for storage of the spent fuel. The fuel itself is usually stored in metallic tubes or canisters to ensure the containment of the radioactive material. Depending on its design, a tube or canister encapsulates one or several assemblies or flasks (cocilles) containing vitrified HLW. The required cooling is provided by natural or forced convection. Examples of the storage of used nuclear fuel in vaults include the modular vault dry storage (MVDS) facility at Paks in Hungary (Huszák and Bencze 2004) or the dry storage vault system at CASCAD in France (Bardelle 1993). In the MVDS, the VVER fuel assemblies are stored vertically in individual storage tubes surrounded by inert

Figure 6.2. The closing of a loaded CASTOR® V/19 cask underwater. Picture taken from GNS (2021a).

nitrogen gas. The decay heat is removed by an ambient air flow that effectively cools the fuel tubes. The design capacity consists of a total of 33 vaults capable of storing approximately 16 000 fuel assemblies (Huszák and Bencze 2004, Hollo and Toth 2012).

Another vault-type facility is the HABOG[4] storage facility in the Netherlands. This facility is designed for the treatment and storage of heat-generating and non-heat-generating HLWs. The heat-generating HLW (e.g. fuel assemblies from research reactors or processed spent fuel from nuclear power plants) are packed in canisters and stored in cylindrical steel wells. The storage wells are equipped with a double jacket and filled with an inert gas to prevent corrosion. Natural convection is used to cool the canisters. A separate compartment of the HABOG facility stores the non-heat-generating HLW (e.g. the residues from fuel cladding and further structural elements of the fuel assembly). The Netherlands has adopted the strategy of storing all HLWs in dedicated surface facilities such as HABOG for at least 100 years (NED 2018).

Storage in silos consists of concrete cylinders that enclose metal liners or metal casks holding the fuel assemblies. The concrete has the safety function of shielding the radiation, and the sealed metal liners or casks ensure the further confinement of stored fuel assemblies. Cooling is generally achieved by natural convection.

---

[4] *Hoog radioaktief Afval Behandelings- en OpslagGebouw*—HABOG.

An example of a silo is the NUHOMS® dry spent fuel storage system, see e.g. Chopra *et al* (2014). It is designed to store PWR and BWR fuel assemblies in dry shielded casks placed horizontally in a concrete storage module. To ensure sufficient protection against neutron and gamma radiation, the concrete structure is approximately 6.1 meters long, 1.7 meters wide and 4.6 meters high, with walls more than 90 cm thick. The structure is modular and the storage system can be extended. The NUHOMS® dry storage cask is a welded cask that basically consists of an internal basket assembly, the shell, and shielding plugs. Depending on the decay power, the cask has a capacity of 24, 32, or 37 PWR fuel assemblies or 52, 61, or 69 BWR fuel assemblies. The heat generated per assembly is limited to 0.3 to 1.2 kW or 18 to 24 kW per canister. Assuming, for example, 24 PWR fuel assemblies with an average burnup of 40 GWd / tHM and a storage (decay) period of 10 years would lead to a heat production of 0.66 kW per fuel assembly or 15.84 kW for the cask. The basket assembly of the NUHOMS® dry cask provides structural support and criticality control. Most basket designs include neutron absorbers made of borated metal components e.g. borated aluminium. The majority of the components are made of stainless steel and aluminium alloys. The dry storage canister, consisting of a rolled and welded stainless steel cylinder, serves as an additional confinement boundary.

The dry storage canister is placed inside the horizontal storage module, which is made of a steel reinforced concrete structure. This structure serves as a biological shield for the radiation and as a protection against natural phenomena, e.g. earthquakes. Even though some design variations exist, the major structural components, apart from the concrete structure, are the access door, the dry storage cask support assembly, and the passive ventilation system. The access door is used to transfer the cask in and out. A steel sleeve is embedded for insertion and retrieval. A thick concrete shielded door covers the access opening. The dry storage cask support system is constructed from steel variants. Rails support the cask position. The passive cooling of the NUHOMS® dry spent fuel system is designed to remove the decay heat from the cask and ensure a peak cladding temperature that does not exceed regulatory limits. The cooling system consists of a shielded air inlet at the lower front. The ventilation air enters the concrete structure via a plenum, then flows around the cask and exits through the roof of the concrete structure. To further prevent the degradation of the concrete, enhanced cooling is achieved by heat shielding plates made e.g. of stainless steel or aluminium placed between the cask and concrete.

6.1.3.2 Casks

Depending on the design, a cask can provide both shielding and containment. If the cask is designed for storage only, it is called a single-purpose cask. If the cask design further includes transport, it is a dual-purpose cask, and if the fuel assemblies remain in the cask even for the final deposit, the cask is a so-called multi-purpose cask. Even within one type, a wide variety of different casks exists. For an overview of different casks, see e.g. Appendix A in Chopra *et al* (2014). In the following, the dual-purpose (transport and storage) cask CASTOR® V manufactured by the *Gesellschaft für Nuklear-Service mbH* (GNS) is described in more detail GNS (2021a) (figure 6.3).

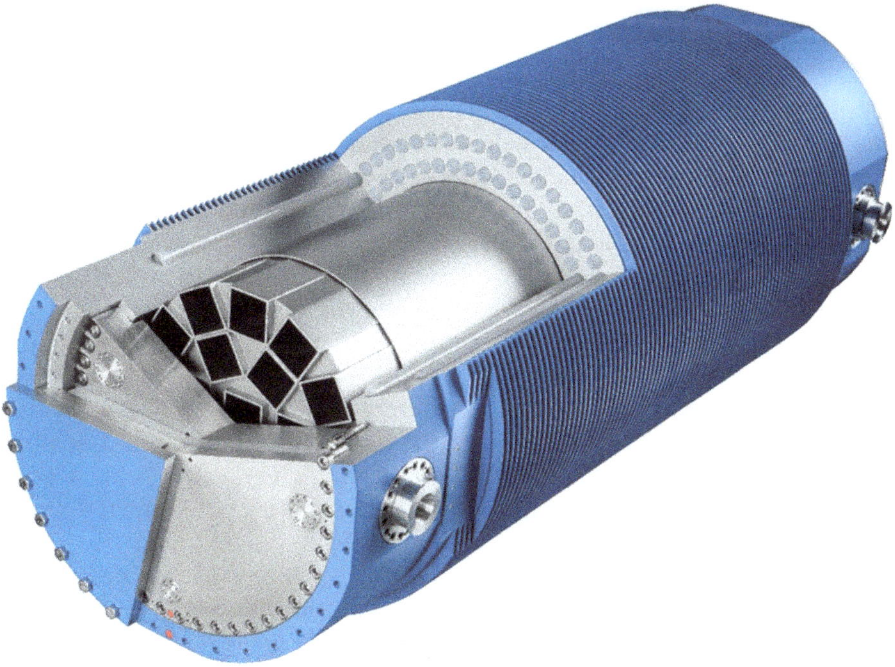

Figure 6.3. CASTOR® V/19 transport and storage cask for PWR fuel assemblies: cutaway view of cask. Picture from GNS (2021b).

The CASTOR® V casks are designed to store fuel assemblies vertically on a concrete pad or inside a storage building. The CASTOR® V/19 cask for PWR fuel assemblies is approximately six meters high in total and has a diameter of 2.44 meters. The height of the cavity is approximately five meters. The walls are approximately 42 cm thick with an additional 6 cm of cooling fins, leaving an inner cavity diameter of approximately 1.5 m. The cask weight is approximately 108 metric tons and, depending on the loading, it weighs around 125 metric tons loaded.

The cask body encloses the basket that holds the fuel assemblies and is closed by a double lid system. The cask is designed and constructed following German design codes and standards, and can withstand e.g. the impact of a one-ton missile hitting the cask with an approximate velocity of 1000 km h^{-1}. It meets the IAEA international specifications for packages of type B(U) (see section 6.2).

The cask body is constructed from a single piece of thick-walled cast iron body with two stainless steel lids bolted to the cask. Neutron and gamma shielding is provided by the thick cask walls and additionally by polyethylene rods inside the cask wall and slabs at the top and bottom of the cask. For better heat transfer and to increase the cooling capacity, circumferential cooling fins surround the cask body. The cask outer shells are covered with epoxy resin to prevent corrosion and facilitate the necessary decontamination, e.g. after the loading of the cask inside the reactor fuel pool[5]. Two upper and lower trunnions allow the cask to be lifted and turned. Two lids made of stainless steel close the cask. Three openings are provided in the primary lid. The first goes straight through the cask and is intended for water filling and draining operations. It is sealed with a shielding cover. Another opening is provided for vacuum drying and helium backfill. The third opening leads to the lower edge of the lid and is intended for leakage testing. The secondary lid above the primary also has three openings. One of them includes the pressure sensor. During cask loading, gas is pumped into the area between the lids at a slightly higher pressure of approximately six bar. A leak in the primary lid would result in a reduction of the pressure in the intra-lid area and the pressure sensor would send an alarm. This would allow the operators to react prior to the escape of radioactive particles from the intra-lid area.

A variety of basket designs exist for different loadings of PWR and BWR fuel assemblies. The design can e.g. depend on the radiation history of the assembly and include additional neutron absorbers and heat transporting components. Generally, the space between the basket and the top of the cask inside is left open for effective circulation of the gas inside, allowing better cooling.

If the casks need to be transported, additional protections can be attached to them, e.g. shock absorbers or impact-limiting components.

6.1.4 Comparison of wet and dry storage

As described above, LWR fuel assemblies are stored in wet storage, namely the fuel pool at the reactor, at least for a shorter period. If the fuel needs to be stored for a longer period, the assemblies can be transferred to either further wet storage away from the reactor or to a dry storage system. Both approaches are well established. Wet storage has the advantage that the fuel assemblies are accessible and visible. However, one disadvantage is the continuous production of additional radioactive waste, e.g. contaminated filters and solutions. The effort needed to operate wet storage is also higher than that for a dry storage facility. Dry stored fuel assemblies are welded or bolted into casks and are thus inaccessible until they are reopened. The inspection of the cask interior has to rely on indirect measurements of e.g. surface heat or gamma profiles. To satisfy the inspection requirement that fuel assemblies remain in the cask once loaded, the IAEA uses a dual seal approach: the sealing and loading is closely monitored and the sealing is regularly inspected. Generally, the requirements for physical security are comparatively lower due to the designs of the cask, vault, or silo, compared to those for wet storage.

However, both wet and dry storage have their advantages and disadvantages. Both are well established and based on decades of operational experience in several countries. A good overview can also be found in IRSN (2019).

[5] Note that for the underwater loading procedure, an additional coat can be applied to the cask to minimize contact with the contaminated water in the fuel pool.

6.1.5 Long-term storage

While final repositories exist for low- and medium-level waste operations, no final disposal of high-level waste has yet reached operational status. Almost all countries that operate nuclear power plants are experiencing a significant delay in finding the end point of the nuclear waste strategy (see part I, chapter 7). Consequently, the HLW accumulates in the form of used fuel assemblies in interim storage facilities, and what was originally planned as an interim storage solution for no more than a few decades has to be prolonged significantly.

The IAEA member states have similar regulatory objectives regarding the management of used nuclear fuel, see table 6.1 for an exemplary overview. A general target of the international licenses for interim storage facilities is around 40 to 50 years. A lifetime extension of the spent fuel storage systems and facilities requires a careful re-examination of the safety requirements (IAEA 2020).

Safety criteria have to be re-examined on a sound scientific–technical basis, which is currently under further international development. This includes the long-term behaviour of spent fuel, fuel assemblies, and packages in both wet and dry storage.

Under normal operations, degradation processes in wet storage are mainly related to the corrosion status of the materials. Certainly, all materials corrode in the wet environment, but for a storage duration of 50–100 years, the corrosion might become a life-limiting factor. The corrosion process can be further affected by the pool water chemistry. However, unlike the case of dry storage, more direct and close monitoring of the material properties is possible and allows a suitable response on the part of the operator, e.g. the encapsulation of single rods which appear to be at risk.

The long-term behaviour of some safety components of the dry storage casks need further investigation. This includes, for example, the sealing of bolted lids and organic neutron shielding.

The continuous change of material properties for both fuel and cladding over the entire storage period needs to be taken into account for the re-examination of safety criteria. It is worth remembering that the spent fuel cladding is the first containment

Table 6.1. Overview of some international AFR storage arrangements, including the individual design life and license durations. The data are taken from IAEA (2002).

Country	Storage	Design life (years)	License duration (years)
Czech Republic	Dual-purpose cask	40 (facility: 50)	10
Germany	Dual-purpose cask	40	40
Korea	Concrete silo	50	Linked to reactor license
Netherlands	Dry storage	100	Not limited
Russia	Wet	30	30
Switzerland	Dual-purpose cask	40	Not limited
United Kingdom	Wet/modular vault	Varies	Not limited
United States	Dry / Wet	40 / linked to reactor	20 / 20

barrier for the radioactive materials and its mechanical strength is the key factor for the operations following dry storage, e.g. further transport. A failure of the cladding could result in a distribution of radioactive material into the cask. Generally, direct inspection or monitoring of the cask interior is impossible without opening the cask. A detailed understanding and investigation of long-term fuel rod behaviour is necessary to predict its integrity for timescales exceeding 40 years. Phenomena that could affect the integrity include e.g. cladding creep, stress corrosion cracking, or delayed hydride cracking. To limit these phenomena, several safety criteria have been established for the originally foreseen storage period. For example, a cladding temperature limit of 370 °C during and after the drying process and storage reduces the cladding degeneration caused by creep. Some regulations foresee further limiting factors on the rod stress and strain. These criteria need to be reevaluated for long-term storage. In particular, the resulting hydrogen behaviour and the formation of hydrides in the cladding needs to be reevaluated under the assumption of lower cladding temperatures as a result of the extended storage (see Bilone *et al* 2013, Kaufholz *et al* 2018, and Motta *et al* 2019). While creep was the most important cladding degeneration mechanism for the initial licensed storage periods of up to 50 years, for extended storage, hydride embrittlement is now seen to be more important (see EPRI 2011, Kook *et al* 2013, and Teague *et al* 2019).

It is worth noting that a loss of fuel rod integrity would affect the retrieval and handling of the fuel assemblies in preparation for the final repository. One of the goals of the spent fuel interim storage is to maintain the retrievability of the fuel assemblies. The definition of retrievability might also depend on the next step foreseen in the waste management strategy. IAEA (2002) suggests that it may be appropriate to define retrievability in terms of the integrity of the storage canister or the basket containing the fuel.

Regardless of the wet and dry storage concepts, once the final repository is available, the fuel assemblies need to be transported from the interim storage facility to the conditioning site.

6.2 Transport of radioactive material

Due to their ability to emit ionizing radiation, radioactive substances are required for a variety of purposes (see chapter 3 'Other waste origins') around the world. In medicine, for example, radioactive substances are used for imaging procedures in diagnostics or within radiation therapy. Radioactive substances are, from a research perspective, useful tools for tracking biochemical processes in organisms or for the development and study of new materials. In industry, the most common areas of application include the calibration of measurement devices, non-destructive testing of materials, and measurements of flow rates, fill levels, or densities. One of the best-known applications is the use of radioactive substances as fuel in nuclear power plants to generate electrical energy.

In the vast majority of cases, the place of manufacture of radioactive substances does not correspond to their place of use, and it is necessary to transport these substances, often over long distances and between countries. The relevance of transport can be illustrated particularly clearly using the example of the nuclear fuel cycle. For reactor operations, it is necessary to ensure a continuous supply of fuel. The required uranium is mined in the form of ores, chemically processed, converted into uranium hexafluoride, enriched, and ultimately processed into fuel assemblies. These fuel assemblies are then transported to the respective reactors. In return, spent fuel assemblies have to be transported to reprocessing facilities or to an interim storage facility. All these steps generally take place at different production sites, and therefore transport is needed.

According to estimates produced by the International Atomic Energy Agency (IAEA), around 20 million transports of radioactive materials are carried out every year. The overwhelming majority of these transports consist of packages with low activity (a few kilo- to gigabecquerels), such as radiopharmaceuticals, measuring devices, or smoke detectors. Transports related to the nuclear fuel cycle represent only a fraction of approximately 5% of the total number (Ratin *et al* 2014).

6.2.1 Statutory basis of the transport of radioactive substances

Radioactive substances pose a wide range of dangers to humans and the environment. Apart from the harm caused by direct radiation exposure, other hazards can arise due to contamination, decay heat or, in the case of fissile materials, unwanted chain reactions. In addition, the chemical compounds themselves can pose other conventional dangers due to their possible caustic or corrosive properties. Radioactive substances are therefore classified as dangerous goods and their transport is subject to special regulations and requirements.

On the advice of the Economic and Social Council of the United Nations (ECOSOC), a committee was commissioned in 1953 to draw up basic recommendations and regulations to ensure the safe transport of dangerous goods. In 1956, this work resulted in the publication of the 'UN Recommendations on the Transport of Dangerous Goods'[6] (UNECE 2017). These recommendations formed the basis of an internally standardized classification and labelling system for dangerous goods and established important standards. Within this system, dangerous goods were divided into classes according to the dangers they posed and were given defined UN numbers and uniform shipping names. Radioactive substances were classified as class 7 (see table 6.2). In addition, minimum requirements for shipping documentation were drawn up and uniform pictograms for quick identification of the dangerous goods classes (without the need to read anything) were developed.

In order to meet the special and industry-specific requirements for radioactive substances, the IAEA was commissioned in 1959 to draw up recommendations for class 7 transports. In 1961, the IAEA published the 'Regulations for the Safe Transport of Radioactive Material (Safety Series No. 6).' These regulations have been revised over the course of time and now include aspects of the following topics (IAEA 2018):

[6] Because of the characteristic orange cover of the recommendations, they are also referred to as the 'Orange Book.'

Table 6.2. List of the nine classes of dangerous goods according to the 'UN Recommendations on the Transport of Dangerous Goods.'

Classes	Description
1	Explosives
2	Gases
3	Flammable liquids
4	Flammable solids
5	Oxidizing agents and organic peroxides
6	Toxic and infectious substances
7	***Radioactive substances***
8	Corrosive substances
9	Miscellaneous dangerous substances and articles, including environmentally hazardous substances

- General provisions (i.e. radiation protection, emergency response, management system, ...)
- Activity limits and classification
- Requirements and controls for transport
- Requirements for radioactive material and for packaging and packages
- Test procedures
- Approval and administrative requirements

Today, they are known as the 'Regulations for the Safe Transport of Radioactive Material,' IAEA Safety Standards Series No. SSR-6 (IAEA 2018). In the course of revisions of the UN recommendations, more and more references were created between the documents, until finally, in 1999, the IAEA Regulations were fully implemented in the UN Recommendations.

From a legal point of view, the UN recommendations are not binding, but they form an important basis for international and national regulations and have thus a legal meaning. Some examples of the most important international agreements for the transport of dangerous goods are:

- IMDG-Code — *International Maritime Dangerous Goods-Code*
  The IMDG code lists the transport regulations for the transport of dangerous goods by sea.

- INF-Code — *International Code for the safe Carriage of Packed Irradiated Nuclear Fuel, Plutonium and High-level Radioactive Wastes on Board Ships*
  The provisions regulate safety requirements for the transport of irradiated fuel assemblies and highly radioactive waste by sea.

- ICAO-TI — *Technical Instructions for the Safe Transport of Dangerous Goods by Air*

The Technical Instructions describe the transport regulations that apply to dangerous goods carried by air.

- RID *Règlement concernant le transport international ferroviaire de marchandises dangereuses*
  The RID describes the regulations for the transport of dangerous goods by rail.

- ADR *Accord relatif au transport international des marchandise Dangereuses par Route*
  The ADR describes the Regulations for the carriage of dangerous goods by road.

Before the international agreements can achieve a legally binding effect on a national basis, they must first be incorporated into the laws and ordinances of the respective countries.

As with the UN recommendations themselves, all other international regulations are refined and revised about every two years. This offers the participating member states the chance to include concerns and take into account the state of the art of science and technology. Regardless of the mode of transport, the common goal of the regulations is to ensure the safety of humans and the environment and to avoid any harm.

6.2.2 Classification and labelling

The classification of radioactive substances is carried out across all modes of transport in accordance with the UN model regulations and the SSR-6 of the IAEA. Radioactive material means any material containing radionuclides in which both the activity concentration and the total activity in the consignment exceed specified values[7] (UNECE 2017). In total, 25 UN numbers[8] are assigned to this class, which can also be assigned to eight different categories. Table 6.3 lists examples of some of the UN numbers in the various categories (in bold in table 6.3). With the help of the defined criteria[9] listed in the relevant regulations, radioactive substances can be assigned to the right category and the corresponding UN number.

In addition to an indication of the UN number and the proper shipping name, further information is relevant for package labelling. In the case of radioactive shipments, information about the consignor or the consignee must be given on the package, in order to be able to make a clear assignment to a contact. In addition to the gross weight[10], the type of package[11] and the name of the manufacturer[12] must be specified. For type B and type C packages, additional information is required,

[7] The values of the radionuclides are specified in the UN Model Regulations.

[8] UN 3507 URANIUM HEXAFLUORIDE, RADIOACTIVE MATERIAL, EXCEPTED PACKAGE is assigned to class 6.1, toxic substances.

[9] Decisive criteria include, among other factors, the fissionability, activity, surface contamination, special form, and the nature of the radioelement.

[10] If the gross mass >50 kg.

[11] That is, types A, B(U), and B(M).

[12] For IP-II, IP-III, and type A.

Table 6.3. Overview of some examples of class 7 UN numbers.

UN No.	Proper shipping and description
	Excepted packages
UN 2910	RADIOACTIVE MATERIAL, EXCEPTED PACKAGE—LIMITED QUANTITY OF MATERIAL
UN 2911	RADIOACTIVE MATERIAL, EXCEPTED PACKAGE—INSTRUMENTS or ARTICLES
	Low specific activity radioactive material
UN 3321	RADIOACTIVE MATERIAL, LOW SPECIFIC ACTIVITY (LSA-II), non-fissile or fissile-excepted
UN 3324	RADIOACTIVE MATERIAL, LOW SPECIFIC ACTIVITY (LSA-II), FISSILE
	Surface contaminated objects
UN 2913	RADIOACTIVE MATERIAL, SURFACE CONTAMINATED OBJECTS (SCO-I OR SCO-II), non-fissile or fissile-excepted
UN 3326	RADIOACTIVE MATERIAL, SURFACE CONTAMINATED OBJECTS (SCO-I OR SCO-II), FISSILE
	Type A packages
UN 2915	RADIOACTIVE MATERIAL, TYPE A PACKAGE, non-special form, non-fissile or fissile-excepted
UN 3332	RADIOACTIVE MATERIAL, TYPE A PACKAGE, SPECIAL FORM, non-fissile or fissile-excepted
	Type B(U) packages
UN 2916	RADIOACTIVE MATERIAL, TYPE B(U) PACKAGE, non-fissile or fissile-excepted
	Type B(M) packages
UN 2917	RADIOACTIVE MATERIAL, TYPE B(M) PACKAGE, non-fissile or fissile-excepted
	Type C packages
UN 3323	RADIOACTIVE MATERIAL, TYPE C PACKAGE, non-fissile or fissile-excepted
	Uranium hexafluoride
UN 2977	RADIOACTIVE MATERIAL, URANIUM HEXAFLUORIDE, FISSILE

such as the type identification, the serial number, and a permanent fire- and waterproof label that includes the ionizing radiation symbol. Depending on the radiological properties, a package can be provided with different hazard labels. The decisive factors in the selection of the labels are the surface dose rate (on the outside of the package), the transport index (TI)[13], and the fissionability (table 6.4).

[13] Obtained by determining the maximum radiation level in units of mSv/h at a distance of 1 m from the external surface of the package, overpack, container, or unpacked LSA-I and SCO-I. The value determined shall be multiplied by 100 and the resulting number is the transport index. For tanks, containers, and unpacked LSA-I and SCO-I, the value might be multiplied by a size-dependent factor.

Table 6.4. Categories of packages, overpacks, and containers.

Conditions			
Transport index (TI)	Maximum radiation level at any point on the external surface (SRL)	Category	
0[a]	SRL $\leqslant 0.005$ mSv h^{-1}	I-White	
$0 < TI \leqslant 1$	0.005 mSv h^{-1} < SRL $\leqslant$ 0.5 mSv h^{-1}	II-Yellow	
$1 < TI \leqslant 10$	0.5 mSv h^{-1} < SRL $\leqslant$ 2 mSv h^{-1}	III-Yellow	
$10 < TI$	2 mSv h^{-1} < SRL $\leqslant$ 10 mSv h^{-1}	III-Yellow[b]	
If fissile materials are transported, this has to be indicated on the package by use of the appropriate hazard label.		Fissile	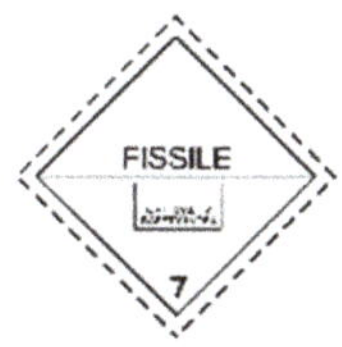

[a] if the TI is no greater than 0.05, the value quoted may be zero
[b] shall be carried under exclusive use

6.2.2.1 Packages and packaging

As a result of its ionizing radiation and the associated harm, the transport of radioactive material is always related to a certain amount of risk for humans and the environment. The regulations for the safe transport of radioactive materials are intended to contain these risks as far as possible and to ensure the optimum protection. This is mainly achieved through the following measures (UNECE 2017):

- Containment of the radioactive contents
- Control of the external dose rate
- Prevention of criticality
- Prevention of damage caused by heat

The implementation of these measures is also achieved through the use of special transport packaging that is manufactured according to defined design requirements and is appropriate to the hazard potential of the contents. This means that the testing requirements for such packaging often increase as the activity inventory increases and verification of the packaging's resistance to various stresses must be provided. The stresses on the packaging also follow a graded approach concerning the conditions of carriage and range from routine transport (incident-free transport) over a normal transport (minor incidents) to transport with accident conditions. The IAEA has defined five different types of package[14] in SSR-6:

- Excepted package
- Industrial package (Type IP-1, Type IP-2, Type IP-3)
- Type A package
- Type B package (Type B(U), Type B(M))
- Type C package[15]

Excepted packages are designed for the transport of radioactive material with very low activity, such as articles manufactured from natural thorium, limited quantities of material, or instruments and articles. Since the harms posed by these contents are considered to be very low, the requirements of the package are designed for incident-free transport. Industrial packages are designed for the transport of materials with higher activity, such as low-specific-activity (LSA) material and surface contaminated objects (SCO). Since the harms posed by these contents are higher, the requirement for solid inclusion of the radioactive material is also higher. Industrial packages have to retain their protective effects not only under incident-free transport situations but for transport under minor incident conditions. This means that a package shall be designed such that if it were subjected to a minor incident[16], it would prevent loss or dispersal of its radioactive contents and an increase of more than 20% in the maximum radiation level at any external surface of the package. For industrial packages, the requirements increases from type IP-1 to IP-3 packages. Type A packages are designed for the enclosure of radioactive material with an activity up to the radionuclide-specific A1/A2 values[17] and must also maintain their protection under normal and routine carriage conditions.

In contrast to the previously mentioned packages, type B packages can contain radioactive material with relatively high activities, such as spent fuel elements or high-level waste. Therefore, these packages are designed to meet the requirement to withstand accident conditions[18] in transit and restrict the loss of their radioactive contents.

[14] In the case of radioactive materials, 'packaging' means the assembly of components necessary to enclose the radioactivity completely, and 'a package' means the packaging with its radioactive contents as presented for carriage.

[15] For carriage by air.

[16] The relevant regulations include various testing procedures for the simulation of certain transport scenarios. The testing procedures for normal conditions of carriage involve a water spray test, a free drop test, a stacking test, and a penetration test.

[17] The A1/A2 values are activity values that were used to determine the activity limits for the requirements of the ADR/RID. A1 means the activity value of special-form radioactive material ('special form' means an indispersible solid radioactive material or a sealed capsule containing radioactive material); A2 means the activity value of radioactive material other than special-form radioactive material.

[18] The tests consist of a mechanical test (three drop tests, *inter alia* up to nine meters on a solid ground), a thermal test (exposure to temperatures up to 800 °C) and a water immersion test (eight hours under a head of water of at least 15 m).

Since type C packages are designed for carriage by air, their design is even more demanding and they need to meet greater requirements[19] than the type B packages.

Packages[20] that carry fissile material additionally need to prove that subcriticality remains guaranteed under accident conditions.

Since the requirements of package types B and C and those for fissile radioactive materials are so demanding, SSR-6 requires the designs to be approved by the competent authorities[21]. The basic requirements to be met for an approval can be found in SSR-6, section VIII 'Approval and administrative requirements'.

Like the rest of the regulations, the requirements for package approval are subject to constant revision. For example, in the current revision of SSR-6, additional requirements for approval have been placed on packages that are used for shipment after storage, such as 'a gap analysis programme describing a systematic procedure for a periodic evaluation of changes of regulations, changes in technical knowledge and changes of the state of the package design during storage' and a 'justification of considerations to ageing mechanisms in the safety analysis and within the proposed operating and maintenance instructions.' The latter aspect addresses the fact that the material of the packaging can be stressed over long periods due to the ionizing radiation of the contents, for example, by corrosion, stress relaxation, structural changes, etc. In addition, there is a general requirement for all packaging and packages that: 'The design of the package shall take into account ageing mechanisms.'

These requirements have a major and far-reaching impact on future transports, including the associated maintenance of packages, in order to maintain the approved state; this is particularly the case for the thousands of packages that are currently in interim storage facilities. The implementation of these demands is currently being coordinated with the competent authorities.

6.2.3 Securing transports

A safe transport is based on several pillars; one of them is safety, another is security. Safety is achieved by the correct application of the regulations and the use of appropriate packaging that prevents dangerous emanation from the freight itself. Nevertheless, there are additional dangers that arise when radioactive or nuclear material falls into the wrong hands. The concerns cover threats including criminals or terrorists who may acquire radioactive materials and potentially build nuclear explosive devices or radiological dispersal devices (dirty bombs). In particular, radioactive materials are considered to be at their most vulnerable during transport

[19] The testing procedures involve several drop tests, a puncture/tearing test, an enhanced thermal test with an exposure time of 60 min to 800 °C and an impact test, in which the specimen shall be subjected to an impact at a velocity of no less than 90 $m\ s^{-1}$.

[20] Type IF (industrial package-fissile), types AF, BF, CF and H(U)/H(M) (specific packages for uranium hexafluoride).

[21] For package designs for which it is not required that a competent authority issues a certificate of approval (type IP, type A), documentary evidence of the compliance of the package design with all applicable requirements is sufficient.

and therefore need to be secured. To respond to these concerns, the IAEA launched the Nuclear Security Series in 2006. This series comprises four extensive sets of publications, consisting of Nuclear Security Fundamentals, Recommendations, Implementation Guides, and Technical Guides (IAEA NSS 2006). These publications give guidance to parties involved in transport as well as to the regulatory bodies of the member states and enable safe and secure transport.

In the area of security, the IAEA proposes a graduated approach that provides a commensurate use of measures, based, among others, on the harms of the contents of a shipment related to the potential consequences of a malicious act. For radioactive material that poses very low potential radiological consequences, such as most of the excepted packages, LSA-I and SCO-I, specific security measures need not be applied. For these materials, the control measurements required by the safety regulations and a prudent management system are considered sufficient.

Radioactive material with a higher activity level and higher radiological hazards requires stricter safety measures. In order to further subdivide these radioactive materials, criteria such as radionuclides and activity can be used (based on A values or D values[22]), leading to the distinction between a basic security level and an enhanced security level. For instance, a basic transport security level can be achieved by taking the following aspects into account:

- Evaluation and exchange of security-relevant information between the parties engaged in the transport
- Protection and control of security-related information
- Trustworthiness determination of the involved personnel
- Written instructions, procedures, and plans
- Security training
- Receiver and carrier authorization
- Identity verification of shipper and carrier
- Communication during the transport
- Appropriate conveyance considerations
- Conveyance inspections
- Package and conveyance security systems
- Monitoring the shipment
- Continuity of security measures
- Receipt verification

The enhanced transport security measures start with the aforementioned aspects and expand them. One major extension is the development of a detailed security plan that involves the collaboration of the shipper, carrier, receiver, and further parties involved in the transport. The plan should include *inter alia* assignments of responsibility, documentation provisions, weak-point analyses, security measures to be implemented (such as training or equipment), procedures for reporting security incidents, protective measures for sensitive information, and measures for

[22] The D-value is the quantity of radioactive material, which, if uncontrolled, could result in the death of an exposed individual or a permanent injury that decreases that person's quality of life.

monitoring the location of the shipment. The parties involved in the transport should cooperate with each other and with the competent authorities in order to exchange information, apply security measures, and respond to security incidents.

All in all, the applied measures of a functioning security system for radioactive materials are designed to serve three basic security functions: detection, delay, and response. When it comes to the transport of nuclear material, the key functions of the security system are expanded to include the aspects of deterrence and assessment. Like radioactive material, nuclear material is categorized into different categories, according to the potential radiological consequences (including the use of nuclear material in a nuclear explosive device) in the case of a malicious act. This also allows graduated measures to be applied.

Regardless of the category of the nuclear material, some common requirements are imposed, such as minimizing the time for which the nuclear material remains in transport and minimizing the number and duration of transfers. The protection required for the material should be consistent with the appropriate category and a transport system should be used that offers physical protection measures appropriate to the threat assessment or design basis threat. Predictable movement schedules should be avoided by varying times and routes, and advance knowledge of the transport details should be limited to the minimum number of persons. For persons involved in the transport, a predetermination of trustworthiness is required. Routes should be used that avoid areas of natural disaster or civil disorder or those with known threats[23], and the packages or conveyances should not be left unattended.

On the basis of the aforementioned common requirements, category-specific measures are applied to guarantee the greatest possible protection of transports.

6.3 Conclusions

As a result of transportation, radioactive materials can readily be made available and can contribute their great benefits to society. The regulations for the safe transport of class 7 dangerous goods are based on international agreements, recommendations, and the regulations of the United Nations. The regular revision cycles help to ensure that the regulations are up to date and take account of the latest developments in science and technology. The fundamental basis for safe transport consists of safety requirements, for example, those for proper packaging and documentation and for security to avoid any malicious misuse of radioactive materials.

Bibliography

Bardelle P 1994 *Dry Vault For Spent Fuel Depository: Basic Outsets, Operating Results and Safety of the "CASCAD" Plant* CEA-CONF-12083

Billone M C, Burtseva T A, Han Z and Liu Y Y 2013 *Used Fuel Disposition Campaign—Embrittlement and DBTT of High-burnup PWR Fuel Cladding Alloys* ANL-13/16; FCRD-UFD-2013-000401(U.S. Dept. of Energy, OSTI)

[23] The complete list of common requirements for the transport of nuclear material can be found in IAEA NSS (2006).

Chopra O K, Diercks D R, Fabian R R, Han Z H and Liu Y Y 2014 *Managing Aging Effects on Dry Cask Storage Systems for Extended Long-term Storage and Transportation of Used Fuel (Rev. 2)* ANL-13/15 107500 (U.S. Dept of Energy, OSTI)

EPRI (Electric Power Research Institute) 2011 *Extended Storage Collaboration Program (ESCP)—Progress Report and Review of Gap Analyses* EPRITR-1022914

GNS 2021a GNS CASTOR® V/19 cask product information www.gns.de/binary.ashx/~default.download/29723 [accessed March 2021]

GNS 2021b GNS Gesellschaft für Nuklear-Service mbH www.gns.de/

Hanson B, Alsaed H, Stockman C, Enos D, Meyer R and Sorenson K 2012 *Gap Analysis to Support Extended Storage of Used Nuclear Fuel* (*Rev. 0*) FCRD-USED-2011-000136 Rev. 0; PNNL-20509

Hollo D and Toth K 2012 *Overview of the Paks Modular Vault Dry Storage Probabilistic Safety Assessment* NEA-CSNI-R--2012-4 (Nuclear Energy Agency of the OECD (NEA))

Huszák Z and Bencze J 2004 Storage of spent nuclear fuel in MVDS of PAKS NPP, dysnai www.dysnai.org/Reports/2000-2004/3.pdf

IAEA (International Atomic Energy Agency) 1996 *Review of Design Approaches of Advanced Pressurized LWRs* IAEA-TECDOC-861 (Vienna: IAEA)

IAEA (International Atomic Energy Agency) 2002 *Long Term Storage of Spent Nuclear Fuel - Survey and Recommendations* IAEA-TECDOC-1293 (Vienna: IAEA)

IAEA (International Atomic Energy Agency) 2006 *Storage of Radioactive Waste* Safety Guide WS-G-6.1 (Vienna: IAEA)

IAEA (International Atomic Energy Agency) 2018 *Regulations for the Safe Transport of Radioactive Material* IAEA Safety Standards Series No. SSR-6 (Rev.1) (Vienna: IAEA)

IAEA (International Atomic Energy Agency) 2020 *Storage of Spent Nuclear Fuel* IAEA Safety Standards Series No. SSG-15 (Rev. 1) (Vienna: IAEA)

IRSN 2019 *Storage of Nuclear Spent Fuel: Concepts and Nuclear Safety Issues* Report No 2019-00181

IAEA NSS 2006 *International Atomic Energy Agency, Nuclear Security Series* https://iaea.org/resources/nuclear-security-series

Kaufholz P, Stuke M, Boldt F and Péridis M 2018 Influence of kinetic effects on terminal solid solubility of hydrogen in zirconium alloys *J. Nucl. Mater.* **510** 277–81

Kook D, Choi J, Kim J and Kim Y 2013 Review of spent fuel integrity evaluation for dry storage *Nuc. Eng. Tech.* **45** 115–24

Motta A T *et al* 2019 Hydrogen in zirconium alloys: a review *J. Nucl. Mater.* **518** 440–60

NED 2018 Radioactive waste management programmes in OECD/NEA member countries: Netherlands, OECD/NEA www.oecdnea.org/upload/docs/application/pdf/2020-07/netherlands_report_web.pdf [accessed May 2021]

OECD 2006 *The Roles of Storage in the Management of Long-lived Radioactive Waste—Practices and Potentialities in OECD Countries* NEA No.6034

OECD 2015 *Status Report on Spent Fuel Pools Under Loss-of-coolant Accident Conditions—Final Report* NEA/CSNI/R(2015)2 (Paris: OECD)

ORANO 2021 www.orano.group/usa/en/our-portfolio-expertise/used-fuel-management/used-fuel-storage [accessed March 2021]

Ratin A S, Mithel M F H, Rishad A S and Nitu T A K 2014 Radioactive waste management of nuclear power plant *Int. J. Renew. Energy Technol.* **3** 1–10

SKB 2021 *CLAB—central interim storage facility for spent nuclear fuel, SKB, Sweden* www.skb.se/publikation/1109049/Clab.pdf [accessed 1 March 2020]

Teague M C, Saltzstein S J, Hanson B, Sorenson K B and Freeze G A 2019 *Gap Analysis to Guide DOE R&D in Supporting Extended Storage and Transportation of Spent Nuclear Fuel: An FY2019 Assessment* SAND-2019-15479R681990; TRN:US2102551

UNECE 2017 *Recommendations on the Transport of Dangerous Goods: Model Regulations (Rev. 20)* ST/SG/AC.10/1/Rev.20 (New York and Geneva: United Nations Economic Commission for Europe)

Vitkova M, Gorinov I and Dacheva D 2006 Safety criteria for wet and dry spent fuel storage *6th Int. Conf. on WWER Fuel Performance, Modelling and Experimental Support (Albena, Bulgaria, 19–23 Sep 2005)* INIS-BG--1112

Chapter 7

Disposal and other conceivable strategy 'endpoints' for different types of waste

Klaus-Jürgen Röhlig

7.1 Various ideas proposed for endpoints

For each waste management strategy, the decisions about the ultimate fate of the materials, i.e. the 'strategy endpoint(s)', are of utmost importance; see part I, chapter 4 'Waste management policy and strategy,' in particular, section 4.2.3. Over the decades, many conceivable endpoints have been proposed for the different types of radioactive waste, but far fewer less have been, or are about to be, implemented. If the endpoint of a material is not some kind of reuse, possibly connected with recycling, and if no further treatment related to its characteristics is foreseen or sensible, one can distinguish between 'dilute and disperse' and 'concentrate and contain' management approaches. In the case of radioactive materials, 'dilute and disperse' approaches can only apply to substances with a sufficiently low content of radioactivity as to allow discharge, which is outside the scope of this book. According to present understanding, such discharges may only be acceptable for 'facilities and activities, [which,] during normal operation, generate gaseous and liquid effluents that contain small amounts of radionuclides that may expose the public and the environment to low levels of radiation. In many cases, the complete prevention of the release of such effluents is technically difficult or extremely costly to achieve. In all cases, the resulting doses to any member of the public must be below established limits' (IAEA 2018a). Note that this applies to planned activities during the normal operation of facilities, not to accidents. It is also important to mention that the discharge issue is controversial. This is particularly the case for the discharges of contaminated water after the Fukushima accidents (METI 2020, IAEA 2020a, Greenpeace 2021)—an activity which can be considered to be 'planned,' but was undertaken as a consequence of an accident.

Thus, only 'concentrate and contain' appears appropriate, although not all of the suggested approaches listed below fall into this category: for example, disposal in

doi:10.1088/978-0-7503-3095-4ch7

outer space does not necessarily belong to this category, but still would (if successfully implemented) emplace the materials outside of the inhabited environment. Also, the injection of liquid waste underground does not ensure containment (which, in itself, is an argument against this option). The same holds for disposal at sea, which was actually considered acceptable for some waste categories in the past (IAEA 1999). Amongst the options currently or formerly considered are according to UK Nirex Ltd (2002), Dutton *et al* (2004), and CoRWM (2006):

- Emplacement of solid or solidified materials in onshore landfill or geo-engineered facilities at or below ground (at different depths),
- Emplacement in facilities which are designed to allow waste retrieval throughout their lifetime (i.e. storage-type facilities, technical approaches for which are discussed in part I, chapter 6 'Storage and transport'),
- Disposal at sea,
- Disposal in outer space,
- Disposal under ice sheets,
- Disposal in subduction zones,
- Disposal in offshore seabed formations,
- Underground injection of liquid (e.g. reprocessing) waste,
- Rock melting, i.e. emplacing especially heat-generating waste deep underground and immobilising the toxic substances by melting and subsequently solidifying the surrounding geologic materials.

In past decades, lists of options such as the above, often complemented by reprocessing and recycling options, e.g. partition and transmutation, were subject to scrutiny by various national and international bodies and commissions, examples being the Swiss Expertengruppe Entsorgungskonzepte für radioaktive Abfälle (EKRA 2000), the British Committee on Radioactive Waste Management (CoRWM 2006), the Canadian Nuclear Waste Management Organization (NWMO 2004, 2005), the U.S. Blue Ribbon Commission on America's Nuclear Future (BRC 2012), the German Kommission Lagerung hoch radioaktiver Abfallstoffe (Kommission 2016), or OECD/NEA's Radioactive Waste Management Committee (RWMC, NEA 1995, 2008). The objectives against which the options were evaluated in these reports and the boundary conditions under which the objectives are to be achieved are those relevant for strategy choices in general; see part I, chapter 4 'Waste management policy and strategy' (see also Dutton *et al* 2004):

- Safety of staff and the general public, security, non-proliferation, radiation protection, environment protection, environmental impact
- Ethical aspects (including inter- and intragenerational justice and responsibility)
- Technical and technological aspects (availability and perspectives of technologies vis-à-vis their potential to address objectives)
- Societal and political aspects (governance, political feasibility, acceptance)
- Costs and financial security
- Practical considerations (e.g. flexibility, minimising transports, …)
- Boundary conditions e.g. from legislation and regulation (which might address aspects of the points above) or from the nuclear energy programme

Note that these issues are interrelated (e.g. safety and environmental protection are ethical issues), apply at different levels (e.g. while safety can be considered an overarching principle, technology availability has to do with practicalities), might evolve over time (which is the case e.g. for safety standards or technology availability), can be actively influenced (e.g. technology availability), and might conflict with each other (e.g. safety at different times in the future). How these issues are weighed and balanced against each other is also important.

An additional consideration might be whether or not an option allows for corrective measures in the case of emergencies or unforeseen circumstances or events, e.g. a kind of reversibility. In particular, it is of interest to know whether or not the waste can be retrieved (retrievability); see NEA (2012).

In the studies quoted above, many of the options for endpoints were discarded. Table 7.1 summarises the options ruled out and the reasons given in these studies. For technical ideas and details regarding these options, we refer to the studies mentioned above as well as to the overview provided by the World Nuclear Association (WNA 2021). Note that none of these options allows for reversibility in the sense mentioned above (see section 7.2 below).

The summary is necessarily a generalization, because the vocabulary and categorisation of the studies varied with regard to both the options and the reasons for ruling them out. Also, not all the studies appreciated all the reasons in the same way. Nevertheless, taken together, they provide a consistent picture. It is also worth mentioning that there was little change concerning the reasoning over the decades during which the studies took place. However, the reasons related to technology, or, more precisely, to a lack of proven technology, could admittedly be seen as a self-fulfilling prophecy: these reasons are brought forward to discard options, which results in a situation in which no or little research and development effort takes place

Table 7.1. Discarded endpoint options and the reasons for ruling them out. Options marked by an asterisk (*) were implemented in the past.

Reasons related to deficiencies concerning…	Safety	Security	Environmental protection	Technology	Costs	International treaties
Disposal at sea*			X			X
Disposal in outer space	X	X	X		X	X
Disposal in or under ice sheets	X	X	X	X		X
Disposal in subduction zones			X	X		X
Disposal in offshore seabed formations			X			X
Underground injection of liquid waste*	X	X	X	X		
Rock melting	X	X	X	X		

for these options. Regarding the column 'international treaties,' it should be mentioned that this column refers to treaties which are unlikely to undergo changes in the foreseeable future.

7.2 Considerations of control, flexibility, retrievability, and passive safety

The remaining options are related to emplacement in various technical or geotechnical facilities, which can be distinguished by their general layout, the emplacement depth, the supervision and maintenance measures they require, the envisaged lifetime, and features concerning retrievability. Facilities with a limited lifetime (i.e. storage facilities) are, by definition, not endpoints in the strict sense of the word, but might still be considered in order to procrastinate final (possibly irreversible) decisions and to ease corrective actions (connected with waste retrieval) in the case of emergencies: This type of emplacement is, by definition and design, retrievable, and, moreover, at some point in time, the waste has to be retrieved. This goes along with the necessity for oversight and to supervise and maintain the facility until retrieval.

Some ideas have been proposed that would preserve this state of maintenance and retrievability, which is associated with the need for regular refurbishment, over an indefinite timeframe: 'Nuclear Guardianship advocates the storage of radioactive material above ground, or just below ground in a monitored, retrievable configuration. When the material is stored *where present and future generations can see it*, the maintenance required for its continual isolation from the environment is more readily facilitated. Nuclear Guardianship acknowledges that this form of responsible care will require ongoing monitoring and maintenance. Because of the uniquely vast temporal nature of radioactive materials, these materials will require routine repackaging in order to ensure their safe accommodation. Continual monitoring and maintenance is and will be necessary. No human-made containment vessel will 'outlive' the radioactive materials they attempt to contain.' (Hibakusha 2015).

Ideas involving this concept, which is also known as 'rolling stewardship', date back over decades; see, for example, the references given by EKRA (2000) or NEA (2013), but they are still being discussed and proposed (CCNR 2013).

IAEA terminology distinguishes (IAEA 2018b):

- Disposal: 'Emplacement of waste in an appropriate facility without the intention of retrieval.'
- Storage: 'The holding of radioactive sources, radioactive material, spent fuel or radioactive waste in a facility that provides for their/its containment, with the intention of retrieval.'

Thus, the rolling guardianship concepts mentioned above could be seen as disposal concepts (as they do not necessarily imply the *intention* of retrieval), but they nevertheless use technical features similar to those of storage, and could be therefore seen as a kind of 'expanded' or 'indefinite' storage. Moreover, they have to address the same challenges as long-term storage, including those described in Dutton *et al* (2004), and NEA (2006, 2020):

- Maintaining the safety-related technical conditions of the inventory, containers, and buildings over the storage timeframe by appropriate ageing management, monitoring, inspection, maintenance, and refurbishment,
- the capability to address risks that arise from routine operation, degradation, and accidents by appropriate mitigation and remediation,
- appropriate record-keeping,
- providing security and safeguarding,
- providing protection from jeopardising natural events,
- addressing legal and regulatory environments that potentially change over time, and
- infrastructure, organisational capability, knowledge transfer, financial security, and the political and societal commitment to ensure all the above.

In contrast, 'classical' disposal concepts attempt to minimise such challenges by striving for a state of 'passive' or 'maintenance-free' safety. This is to say that active measures such as monitoring are not excluded, but safety should, at least from a certain point in time onward, not depend on them. This is also to say that retrieval of the materials is not necessarily excluded, but certainly not *intended.* The IAEA recommends a so-called graded approach to disposal, in which 'the ability of the chosen disposal system to provide containment of the waste and to isolate it from people and the environment will be commensurate with the hazard potential of the waste' (IAEA 2011). The choice of disposal facilities thus depends on the characteristics of the waste to be disposed of; possibilities include landfills and engineered facilities above or at varying depths below ground (see below).

The discourse about the degrees of oversight and active control necessary is orientated by safety/security and fairness considerations (Röhlig *et al* 2019, Röhlig 2019). Moreover, arguments and emotions such as those about unduly or 'unethically abandoning' the waste or of 'sweeping it under the carpet' might play a role.

Many fairness considerations address the issue of intergenerational fairness: while proponents of guardianship argue that such solutions provide the maximum flexibility for future generations (even for re-use of the materials), opponents favour disposal options which 'aim to avoid imposing undue burdens on future generations' (IAEA 1997) in that they do not require their attention, activity, or resources. It has to be acknowledged that these concepts of fairness differ considerably from each other (Dutton *et al* 2004, pp 34 ff.). Furthermore, it can be argued that some might prefer flexibility for future generations regarding the management of the waste, while others could argue that such flexibility requires resources, which then, in turn, will lack when striving for flexibility in other areas: 'The focus is on the question […] of whether freedom to act means carefreeness and easing the burden, or rather the freedom to decide on one's own about how to handle the waste.' (Eckhardt 2015, translation by the author of this chapter).

Intragenerational fairness is an issue e.g. when it comes to site selection, but not so much of an issue when considering strategy endpoints at a generic level.

Concerning long-term safety, the concept of 'passive safety' relies on the understanding that the abovementioned challenges concerning long-term storage cannot be met for the timeframes of interest due to the instability and unpredictability of economic and societal structures. In contrast, engineered components and, in particular, geologic formations might be able to offer stability and performance as barriers for much longer times. Various technical disposal concepts have been developed in the past for the different types of radioactive waste. They differ from each other in terms of layout, safety concept, location at or below the Earth's surface, and with respect to the balance between passive safety and active oversight. This balance might evolve over time. In particular, for the disposal of HLW and SNF declared as waste at depths of several hundred meters ('deep' or 'geologic' disposal), several of these concepts involve means of allowing or easing possible retrieval or recovery actions ('retrievability'), although, according to the definition of disposal, such retrieval is not intended. By doing so, these concepts partly attempt to bridge the conflict between the approaches of 'guardianship' and of 'passive safety' described above. A target conflict might exist, though, between the aims of maximum barrier performance and of accessibility. Often, retrievability is the objective for a limited period of time (decades or about 100 years), during which the disposal facility will not (yet) be completely closed, and the ease of potential retrieval will decrease over this period (figure 7.1).

The French concept of HLW disposal at the Bure site is a technically advanced example in which this approach of evolving passive safety has been implemented (Andra 2016).

It should also be noted, though, that several, if not most, deep disposal projects for HLW and SNF (with the exceptions of the French, Swedish, and Finnish projects) will take further decades of research, development, and implementation before the first waste package might be emplaced. It appears therefore advisable to focus flexibility considerations not only on the time after emplacement, but also very much on the time until then. This corresponds with the wider notion of reversibility of actions in a process, retrievability being the specific possibility of reversing the action of waste emplacement in a disposal facility (NEA 2012, figure 7.2).

Waste management plans in which this principle has been implemented include the Canadian Adaptive Phased Management approach (NWMO 2005) and the German siting process for HLW/SNF disposal. For the latter, see part II chapter 16 'The German case for dealing with high-level radioactive waste: taking a socio-technical approach to address a socio-technical problem—chances and risks'. Canadian Adaptive Phased Management consists of both a technical method and a management system. The key attributes of this approach are:

- Ultimate centralized containment and isolation of used nuclear fuel in an appropriate geological formation;
- Phased and adaptive decision-making;
- Optional shallow storage at the central site prior to placement in the repository;
- Continuous monitoring;
- Provision for retrievability; and
- Citizen engagement.

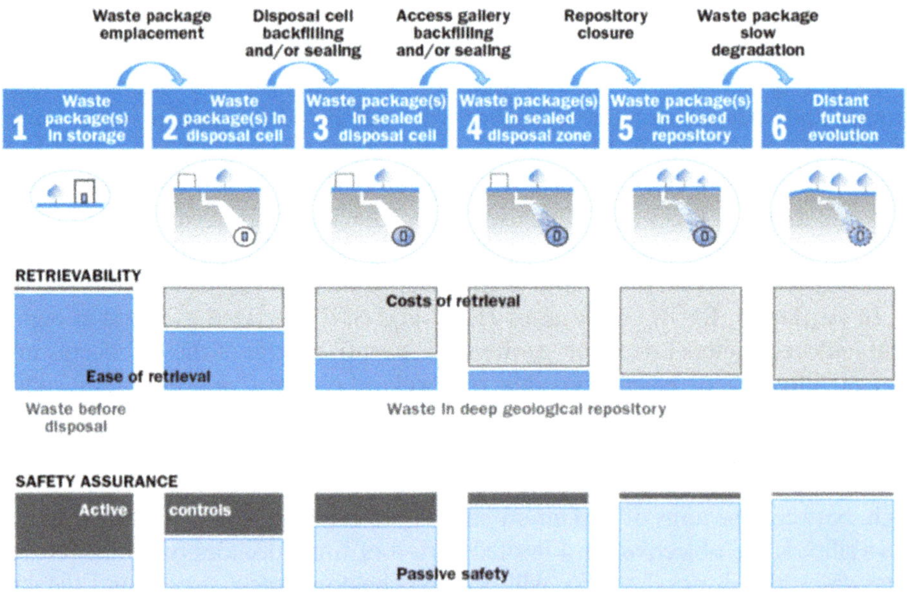

1. During the operational phase, not all waste packages present in the facility will be at the same life cycle stage.
2. Exact proportions of illustrated rectangles may vary depending on the repository design.

Figure 7.1. International 'R-scale': waste life-cycle stages, illustrating changing degrees of retrievability, passive vs. active controls, and the costs of retrieval from a deep geological repository, reproduced by permission of the OECD from NEA (2012). Note that the term 'passive control' stems from the original document—it refers to passive safety.

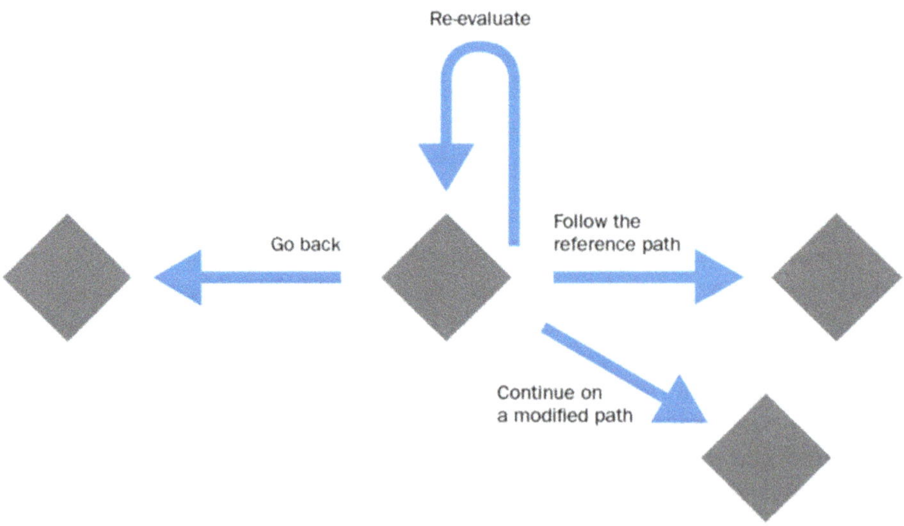

Figure 7.2. Reversibility of decisions—potential outcomes of option assessment, including reversal. Reproduced by permission of the OECD from NEA (2012).

NWMO's plan is directed at disposal in a deep geological repository but flexible with respect to the way towards this goal. A well-defined siting process connected with an engagement program with potentially affected people and communities is a central element of the process (cf. chapter 13 'Role of stakeholder involvement in the implementation of radioactive waste management projects', section 13.4.1). The process includes, inter alia, expressions of interest by communities at various stages as well as their involvement in assessments addressing technical and societal issues important for siting decisions and in site investigation programs. Flexibility measures include 'potential to retrieve it for an extended time. There is provision for contingencies, such as the optional step of shallow storage at the selected central site if circumstances favour early centralization of the spent fuel before the repository is ready.' (CNSC 2017). The process is also flexible with regard to its timing. For details cf. https://www.nwmo.ca/en/Site-selection/Steps-in-the-Process/ [accessed 19 January 2022].

7.3 Technical solutions for disposal—the graded approach

General advice about applying the graded approach to disposal is given by IAEA (2009); see figure 7.3.

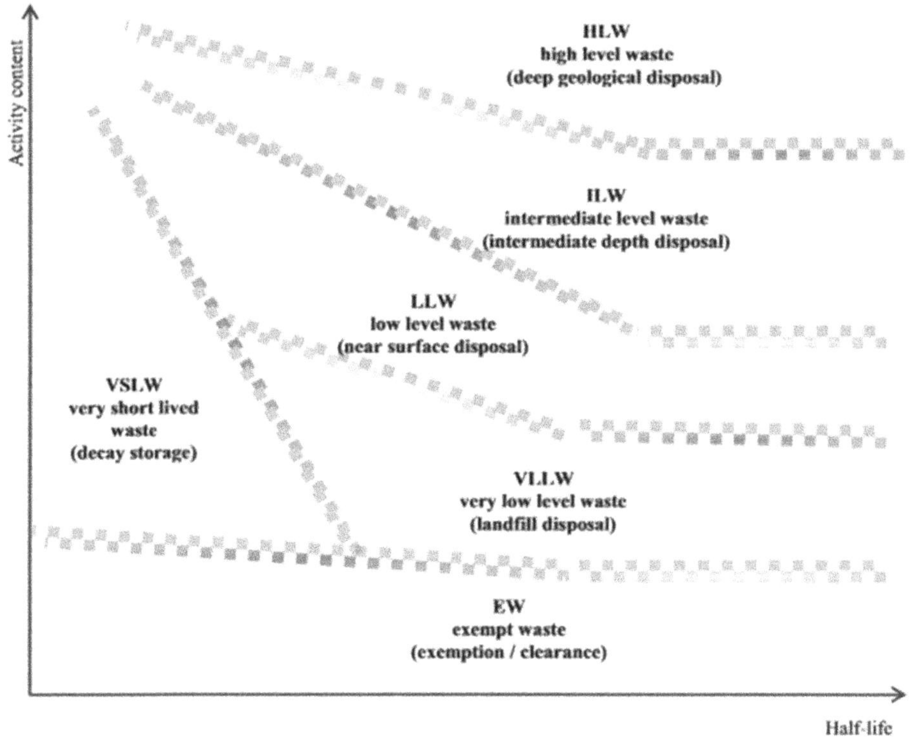

Figure 7.3. Conceptual illustration of IAEA's waste classification scheme, reproduced by permission of the IAEA from IAEA (2009).

As to be expected, the advice is to increase the depth (some tens of meters for intermediate-depth disposal and some hundreds of meters for deep geological disposal) and barrier strength with increasing hazard potential of the inventory. Interestingly, IAEA (2009) establishes this link by a logic of waste categorisation which might appear to be circular reasoning, e.g.: 'High level waste is defined to be waste that contains such large concentrations of both short and long-lived radionuclides that, compared to ILW, a greater degree of containment and isolation from the accessible environment is needed to ensure long-term safety. Such containment and isolation is usually provided by the integrity and stability of deep geological disposal, with engineered barriers.' Similar definitions, or rather descriptions, are given for the other waste types. There are reasons not to be more precise and for the classification scheme in figure 7.3 to be provided without numbers on the axes for the half-life and the activity content: disposal facilities are being planned and designed for specific waste types, but in the end it is necessary to demonstrate for which types a facility is feasible and safe. The means used to link the two issues (waste types versus facility feasibility and safety) are the waste acceptance criteria (WAC); see part I, chapter 4 'Waste management policy and strategy', in particular, 4.2.3. As explained there, WAC work both ways: they are derived in order to ensure disposal feasibility and safety and therefore, *inter alia*, based on a *facility-specific* safety assessment (while, in contrast, IAEA's scheme is not facility-specific but generic). Conversely, it has to be ensured that the set of WAC for all disposal facilities planned or operated within a programme covers the whole waste inventory to be managed by means of the programme. Appropriate conditioning is an important means to achieve this goal.

The disposal options given in figure 7.3 can be further distinguished (IAEA 2017):

- Landfill sites
- Near-surface facilities:
    - Simple near-surface facilities (trenches)
    - Engineered near-surface facilities
    - Near-surface borehole or shaft facilities
- Intermediate-depth facilities
    - Intermediate-depth shafts or boreholes without engineered barrier systems (EBSs)
    - Intermediate-depth shafts or boreholes with EBSs
    - Intermediate-depth repositories
- Deep facilities
    - Deep boreholes without EBSs
    - Deep boreholes with EBSs
    - Mined geological repositories

A schematic illustration of some of these options is given in figure 7.4.

Table 7.2 provides examples of disposal projects but is, in particular for the large number of existing surface and near-surface facilities around the world, by no means comprehensive or complete. Rather, it aims to illustrate the variation and diversity of approaches. Note that there are also facilities which cannot be easily categorised

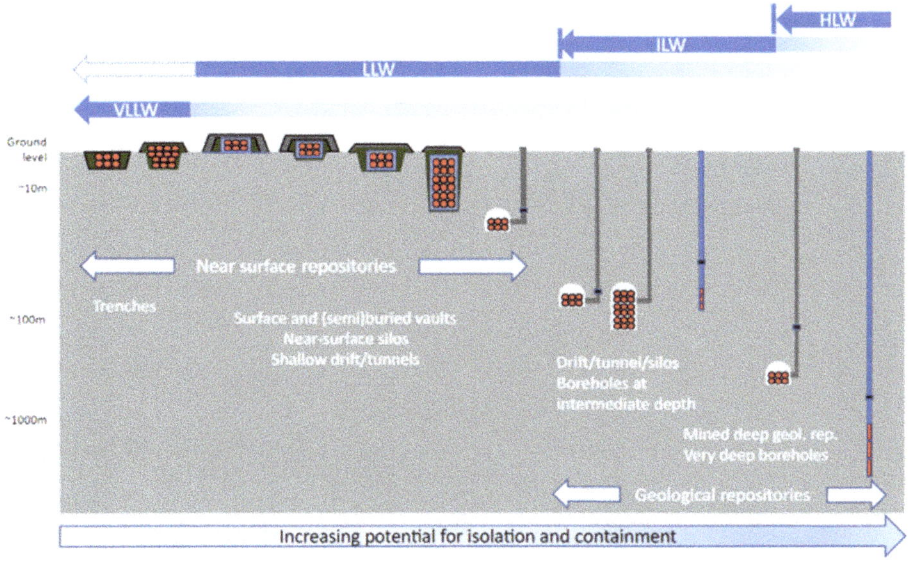

Figure 7.4. Schematic illustration of the range of disposal options, from surface to deep, currently considered or implemented for different classes of radioactive waste, reproduced by permission of the IAEA from IAEA (2020b).

according to figure 7.4, e.g. the former Richard mine (Czech Republic) and the Băiţa-Bihor mine (Romania): the emplacement vaults are accessed by horizontal drifts, therefore it is not clear how the value 'repository depth' should be measured. For many (near-) surface disposal facilities, the safety concept relies on a prolonged monitoring (oversight) period (e.g. 300 years in the case of CSA), during which most of the radionuclides disposed of will decay.

The concept of boreholes near the surface or at intermediate depths has been proposed by the IAEA and the South African Nuclear Energy Corporation (NECSA), in particular for the disposal of disused sealed radioactive sources (DSRSs). Since DSRSs are generally unsuitable for surface disposal, they might be disposed of together with other materials in facilities deeper underground ('co-disposal'). This is infeasible, though, for countries which will never implement such facilities due to the lack of other waste to be disposed of there, i.e. countries without a nuclear programme. 'Disposing DSRS inventories in boreholes was proposed as a possible cost effective solution. Since 1995, the idea has evolved into a well defined concept offering an internationally accepted solution for a wide spectrum of DSRSs for implementation in different geological environments and climatic conditions.' (IAEA 2020b). Currently—with the support of Canada—Ghana, the Philippines, and Malaysia aim to implement such a solution (Cochran *et al* 2018).

Disposal at intermediate depths (i.e. several tens of meters) in tunnels and/or silos takes place e.g. at Loviisa and Olkoluoto (Finland, STUK 2020, TVO 2018), and in Forsmark (Sweden, SKB 2021).

Table 7.2. Some examples of disposal facilities of different types and for different waste categories.

Type	Facility	Waste type (quoted or paraphrased according to national classification and terminology)	Country	References
Near surface, trenches	Centre industriel de regroupement, d'entreposage et de stockage (CIRES) Morvilliers	Very low-level waste	France	Andra (2018a)
Near surface, engineered	Centre de stockage de l'Aube (CSA, department of Aube)	Short-lived low- and intermediate-level waste	France	Andra (2018b)
	Centre de stockage de la Manche (CSM, near La Hague)	Short-lived low- and intermediate-level waste	France	Andra (2018c)
	Dukovany repository	Low-level waste from nuclear power plant operations	Czech Republic	SÚRAO (2019a)
	National Radioactive Waste Repository Mochovce	Solid and solidified low-activity waste from the operation and decommissioning of nuclear power plants, from research institutes, laboratories, and hospitals	Slovakia	JAVYS (2019)
	El Cabril Disposal Facility	Very low-, low-, and intermediate-level waste from hospitals, research centres, industries, and nuclear power plants (very low-level waste in trenches)	Spain	ENRESA (2019)
Near surface, various disposal technologies	Low-Level Waste Repository at Drigg, Cumbria	Low-level solid waste from a range of customers, such as the nuclear industry, the Ministry of Defence, non-nuclear industries, educational, medical, and research establishments	UK	LLWR (2021), ONR (2021)
Former limestone mine, access via horizontal drifts	Richard repository near Litoměřice	Institutional waste from healthcare, industrial, agricultural, and research sectors	Czech Republic	SÚRAO (2019b)

(Continued)

Table 7.2. (*Continued*)

Type	Facility	Waste type (quoted or paraphrased according to national classification and terminology)	Country	References
Former uranium mine, access via horizontal drifts	National Repository for Low- and Intermediate-Level Radioactive Waste—Băiţa-Bihor	Low- and intermediate-level radioactive waste	Romania	CNCAN (2018), Biurrun *et al* (2009)
Intermediate depth, tunnels	Loviisa VLJ repository	Low- and intermediate-level waste generated at the Loviisa power plant	Finland	STUK (2020), TVO (2018)
Intermediate depth, silos	Olkiluoto VLJ repository	Low- and intermediate-level waste generated in state-run industry, medical care, and research; in the future also waste from the Olkiluoto nuclear power plant	Finland	STUK (2020), TVO (2018)
Intermediate depth, vaults and silos	SFR Forsmark	Low- and medium-level waste from the the operations of Swedish nuclear power plants but also from hospitals, veterinary medicine, research, and industry	Sweden	SKB (2021)
Intermediate depth, boreholes		Disused sealed radioactive sources (DSRSs)	Plans in Ghana, the Philippines, Malaysia, Brazil	Cochran *et al* (2018), Cota *et al* (2018)
Deep geologic disposal in mined facilities	Waste Isolation Pilot Plant (WIPP, Carlsbad, New Mexico)	Transuranic (TRU) waste (clothing, tools, rags, residues, debris, soil, and other items contaminated with small amounts of plutonium and other man-made radioactive elements), categorized as 'contact-handled' or 'remote-handled' based on the amount of radiation dose measured at the surface of the waste container	USA	WIPP (2021)

	Asse II (former salt and potassium mine, past practise, waste to be retrieved)	Waste with negligible heat generation	Germany	BGE (2021a)
	ERAM (Morsleben) (former salt and potassium mine, past practise, facility to be closed)	Waste with negligible heat generation	Germany	BGE (2021b)
	Konrad (former iron ore mine, under construction including new emplacement vaults)	Waste with negligible heat generation	Germany	BGE (2021c)
	Several in different stages of conceptualisation, siting, licensing, or construction	Spent nuclear fuel and/or high-level waste from reprocessing, sometimes along with other wastes	Various	Chapter 8 'Geologic ('deep') disposal of high-level and other long-lived waste: host rocks, concepts, current international status'
Very deep boreholes	No site-related projects so far	Various	Various	Rocher and Železnik (2020)

Disposal at depths of at least several hundred meters ('deep geological disposal') 'represents the safest and most sustainable option as the endpoint of the management of high-level waste and spent fuel considered as waste' (EU 2011, preamble). Broadly, two types of deep geological disposal facility can be distinguished:

- Disposal in very (i.e. several thousand meters) deep boreholes, and
- Disposal in deep mined facilities, usually several hundred meters deep.

Very deep borehole disposal has been studied in several countries, but so far, there are no projects that seriously aim to implement it. Its outstanding features are its great disposal depth as well as the fact that all operations (construction, waste emplacement, and closure) would be carried out by staff acting from above ground. Intuitively, the relatively greater depth suggests a higher safety level, but this would need to be confirmed by means of site characterisation, which presents a challenge at such depths. Further challenges are related to drilling, emplacement, and closure technology. These challenges will increase with increasing borehole diameters. Therefore, very deep borehole disposal is sometimes only considered for specific waste types in containers with reasonably small diameters (up to some decimetres). Waste retrieval, if required, is considered to be more challenging than in the case of mined repositories. A recent overview of studies that consider this disposal type has been authored by Rocher and Železnik (2020).

In contrast, disposal in mined facilities takes place at lesser depths and relies on staff working underground during construction, waste emplacement, and closure. Access via shafts and/or ramps allows for the emplacement of containers with varying dimensions and masses (up to tens of tons). This concept relies on passive safety after facility closure, but might allow for retrievability (see above). Several countries aim to dispose of their high-level reprocessing waste and/or their spent nuclear fuel considered as waste in deep mined facilities; as a rule, the identification of a disposal site which is both technically appropriate and publicly accepted is a particular challenge. Three countries (Finland, Sweden, and France) plan to start waste emplacement within this decade. There are also mined disposal facilities for other waste types, which were, are, or are planned to become operational: at the Waste Isolation Pilot Plant (WIPP, New Mexico, USA) defense-generated radioactive waste is being disposed of in a bedded salt layer at a depth of about 600 m (WIPP 2021). In Germany, past practices have included LILW disposal at the Asse II and Morsleben mines (BGE 2021a, 2021b). In the future, it is planned to dispose of LILW at Konrad (BGE 2021c). In Canada, deep LILW disposal at the Bruce nuclear site in Kincardine (Ontario) was proposed by Ontario Power Generation (OPG) in 2002, but recently rejected by the Saugeen Ojibway Nation members (OPG 2021). For more details about deep geological disposal in mined facilities, see the following chapter 8 'Geologic ('deep') disposal of high-level and other long-lived waste: host rocks, concepts, current international status'.

A specific consideration is whether (and if so, to what extent) other waste types (i.e. LILW) should be also disposed of ('co-disposal') at disposal facilities designed for HLW and/or SNF. Various issues need to be taken into account when making such a decision:

- Operational and logistical issues (high variations in waste volumes, container masses, shielding, and cooling requirements),
- Thermal and chemical interactions between the different wastes emplaced, causing a potential safety impact after facility closure,
- Timing of waste arisings versus operational timetable for disposal,
- Governance and acceptance considerations, and
- Economics.

In summary, it can be said that the planned endpoint for spent nuclear fuel and high-level waste in those countries which have already specific plans is deep (geologic) disposal in mined facilities specifically constructed for this purpose. However, no such facility is already operational (see chapter 8 'Geologic ('deep') disposal of high-level and other long-lived waste: host rocks, concepts, current international status'). Note that in addition to countries at different stages of the conceptualisation, siting, licensing, or construction of such facilities, several countries have been using, are using, or plan to use nuclear power without advanced plans for endpoints for the resulting waste.

A much more heterogeneous situation can be observed with regard to other waste types: differences in waste categorisation, in the geologic situation, as well as history (past practises, existing mines etc.) are reasons for this situation.

Bibliography

Andra (Agence nationale pour la gestion des déchets radioactifs) 2016 *Dossier d'Options Techniques de Récupérabilité (DORec)* CG-TE-D-NTE-AMOA-RV0-0000-15–0059/A https://andra.fr/sites/default/files/2018-11/CG-TE-D-NTE-AMOA-RV0-0000-15-0059-A%20DOREC.pdf

Andra (Agence nationale pour la gestion des déchets radioactifs) 2018a *The CIRES. Providing Safe and Sustainable Solutions for VLLW and Non-electronuclear Waste* https://international.andra.fr/operational-facilities/industrial-facility-grouping-sorting-and-very-low-level-waste-vllw-disposal-facility-cires/cires

Andra (Agence nationale pour la gestion des déchets radioactifs) 2018b *The Aube Waste Disposal Facility (CSA). Concept and Design* https://international.andra.fr/operational-facilities/aube-waste-disposal-facility-csa/concept-and-design

Andra (Agence nationale pour la gestion des déchets radioactifs) 2018c *Manche Disposal Facility (CSM)* https://international.andra.fr/existing-facilities/manche-disposal-facility-csm

BGE (Bundesgesellschaft für Endlagerung mbH. Federal Company for Radioactive Waste Disposal Ltd) 2021a *Asse II mine* https://bge.de/en/asse/

BGE (Bundesgesellschaft für Endlagerung mbH. Federal Company for Radioactive Waste Disposal Ltd) 2021b *Morsleben Repository* https://bge.de/en/morsleben/

BGE (Bundesgesellschaft für Endlagerung mbH. Federal Company for Radioactive Waste Disposal Ltd) 2021c *Konrad Repository* https://bge.de/en/konrad/

Biurrun E, Haverkamp B and Röhlig K-J 2009 Regulatory review of preliminary safety assessment for the Băiţa-Bihor Repository, Romania *ASME 2009 12th Int. Conf. on Environmental Remediation and Radioactive Waste Management* https://researchgate.net/publication/267577057_Regulatory_Review_of_Preliminary_Safety_Assessment_for_the_Baita-Bihor_Repository_Romania

BRC (Blue Ribbon Commission on America's Nuclear Future) 2012 *Report to the Secretary of Energy* https://energy.gov/ne/downloads/blue-ribbon-commission-americas-nuclear-future-report-secretary-energy

CCNR (Canadian Coalition for Nuclear Responsibility) 2013 *Nuclear Waste: Abandonment versus Rolling Stewardship* http://ccnr.org/Rolling_Stewardship.pdf

CNCAN (Comisia Naţională pentru Controlul Activităţilor Nucleare. National Commission for Nuclear Activities Control) 2018 *ROMANIA. Joint Convention on the Safety of Spent Fuel Management and on the Safety of Radioactive Waste Management* The Sixth National Report https://iaea.org/sites/default/files/national_report_of_romania_for_the_6th_review_meeting__english.pdf

CNSC (Canadian Nuclear Safety Commission) 2017 *Canadian National Report for the Joint Convention on the Safety of Spent Fuel Management and on the Safety of Radioactive Waste Management* https://www.iaea.org/sites/default/files/6rm-canada.pdf

Cochran J R, Bennett D G, Degnan P, Grout C, Liebenberg G, Little R, Ramsey J, van Blerk J and Van Marcke P 2018 International implementation of IAEA's borehole disposal concept for sealed radioactive sources *WM2018 Conf., (18–22 March 2018, Phoenix, Arizona, USA)* https://researchgate.net/publication/324747105_Int._Implementation_of_IAEA%27s_Borehole_Disposal_Concept_for_Sealed_Radioactive_Sources

CoRWM (Committee on Radioactive Waste Management) 2006 *Managing our Radioactive Waste Safely. CoRWM's Recommendations to Government* https://gov.uk/government/uploads/system/uploads/attachment_data/file/294118/700_-_CoRWM_July_2006_Recommendations_to_Government_pdf.pdf

Cota S, Hirimoto G, Gharbieh H and Silva A 2018 Preliminary post-closure safety assessment for a borehole-type repository for disused sealed radioactive sources in Brazil *Prog. Nucl. Energy* **103** 74–80

Dutton M *et al* 2004 *The Comparison of Alternative Waste Management Strategies for Long-Lived Radioactive Wastes (COMPAS Project)* EUR 21021 EN, CEC, Luxembourg https://op.europa.eu/en/publication-detail/-/publication/c5f948f4-7c05-4f94-8bdf-66342864d973

Eckhardt A 2015 *Stellungnahme zur Anhörung „Sicherheitsanforderungen des BMU 2010' (in German, Expert Opinion given at the Hearing '2010 BMU Safety Requirements')* 19 November 2015, Kommission Lagerung hoch radioaktiver Abfallstoffe https://bundestag.de/resource/blob/395526/498cdb3c87b46b768ce7d1c9b4f4caf3/drs_135-b-data.pdf

EKRA (Expertengruppe Entsorgungskonzepte für radioaktive Abfälle) 2000 *Entsorgungskonzepte für radioaktive Abfälle, Schlussbericht (Disposal concepts for radioactive wastes, Final Report, in German)* https://ensi.ch/fr/wp-content/uploads/sites/4/2012/05/ekra-bericht_entsorgungskonzeptschweiz.pdf

ENRESA (Empresa Nacional de Residuos Radiactivos S.A. National Radioactive Waste Company) 2019 *El Cabril Disposal Facility* https://enresa.es/eng/index/activities-and-projects/el-cabril/

EU (European Union) 2011 Council Directive 2011/70/EURATOM of 19 July 2011 establishing a Community framework for the responsible and safe management of spent fuel and radioactive waste *Off. J. EU* L 199/48 https://eur-lex.europa.eu/legal-content/EN/TXT/?uri=celex%3A32011L0070

Greenpeace 2021 *The Japanese Government's Decision to Discharge Fukushima Contaminated Water Ignores Human Rights and International Maritime Law* https://greenpeace.org/international/press-release/47207/the-japanese-governments-decision-to-discharge-fukushima-contaminated-water-ignores-human-rights-and-international-maritime-law/

Hibakusha 2015 *Nuclear Guardianship* https://hibakushastories.org/why-disarmament-education/nuclear-guardianship/

IAEA (International Atomic Energy Agency) 1997 *Joint Convention on the Safety of Spent Fuel Management and on the Safety of Radioactive Waste Management* https://iaea.org/sites/default/files/infcirc546.pdf

IAEA (International Atomic Energy Agency) 1999 *Inventory of Radioactive Waste Disposals at Sea* IAEA-TECDOC-1105 https://www-pub.iaea.org/MTCD/Publications/PDF/te_1105_prn.pdf

IAEA (International Atomic Energy Agency) 2009 *Classification of Radioactive Waste* General safety guide, IAEA Safety Standards Series No. GSG-1 https://www-pub.iaea.org/MTCD/Publications/PDF/Pub1419_web.pdf

IAEA (International Atomic Energy Agency) 2011 *Disposal of Radioactive Waste* Specific Safety Requirements, IAEA Safety Standards Series No. SSR-5 https://www-pub.iaea.org/MTCD/Publications/PDF/Pub1449_web.pdf

IAEA (International Atomic Energy Agency) 2017 *Selection of Technical Solutions for the Management of Radioactive Waste* IAEA-TECDOC-1817 https://www-pub.iaea.org/MTCD/Publications/PDF/TE-1817_web.pdf

IAEA (International Atomic Energy Agency) 2018a *Regulatory Control of Radioactive Discharges to the Environment*. General safety guide, IAEA Safety Standards Series No. GSG-9 https://www-pub.iaea.org/MTCD/Publications/PDF/PUB1818_web.pdf

IAEA (International Atomic Energy Agency) 2018b IAEA Safety Glossary. Terminology used in Nuclear Safety and Radiation Protection. 2018 edn https://www-pub.iaea.org/MTCD/Publications/PDF/PUB1830_web.pdf

IAEA (International Atomic Energy Agency) 2020a *IAEA Follow-up Review of Progress Made on Management of Alps Treated Water and the Report of the Subcommittee on Handling of ALPS Treated Water at TEPCO's Fukushima Daiichi Nuclear Power Station* Review report https://iaea.org/sites/default/files/20/04/review-report-020420.pdf

IAEA (International Atomic Energy Agency) 2020b *Design Principles and Approaches for Radioactive Waste Repositories* IAEA Nuclear Energy Series No. NW-T-1.27 https://www-pub.iaea.org/MTCD/Publications/PDF/PUB1908_web.pdf

JAVYS (Jadrová a vyraáovacia spoločnosť. Nuclear and Decommissioning Company) 2019 *National Radioactive Waste Repository* https://javys.sk/en/nuclear-facilities/national-radio-active-waste-repository/ru-rao/

Kommision (Kommission Lagerung hoch radioaktiver Abfallstoffe) 2016 *Abschlussbericht der Kommission Lagerung hoch radioaktiver Abfallstoffe. Verantwortung für die Zukunft. Ein faires und transparentes Verfahren für die Auswahl eines nationalen Endlagerstandortes (Final Report of the Commission on the Disposal of Highly Radioactive Waste Substances. Responsibility for the Future. A Fair and Transparent Procedure for the Selection of a National Repository Site, in German)* Deutscher Bundestag 18. Wahlperiode, Drucksache 18/9100 https://dipbt.bundestag.de/doc/btd/18/091/1809100.pdf

LLWR (LLW Repository Ltd) 2021 *About Us* https://gov.uk/government/organisations/low-level-waste-repository-ltd/about

METI (Ministry of Economy, Trade and Industry, Japan) 2020 *The Subcommittee on Handling of the ALPS Treated Water* Provisional translation https://meti.go.jp/english/earthquake/nuclear/decommissioning/pdf/20200210_alps.pdf

NEA (Nuclear Energy Agency) 1995 *The Environmental and Ethical Basis of Geological Disposal of Long-lived Radioactive Wastes. A Collective Opinion of the NEA Radioactive Waste*

Management Committee OECD https://oecd-nea.org/jcms/pl_34631/the-environmental-and-ethical-basis-of-geological-disposal-of-long-lived-radioactive-wastes

NEA (Nuclear Energy Agency) 2006 *The Roles of Storage in the Management of Long-lived Radioactive Waste. Practices and Potentialities in OECD Countries* OECD, NEA No. 6043 https://oecd-nea.org/jcms/pl_14068/the-roles-of-storage-in-the-management-of-long-lived-radioactive-waste

NEA (Nuclear Energy Agency) 2008 *Moving Forward with Geological Disposal of Radioactive Waste: An NEA RWMC Collective Statement* OECD, NEA No. 6433 https://oecd-nea.org/jcms/pl_14450/moving-forward-with-geological-disposal-of-radioactive-waste

NEA (Nuclear Energy Agency) 2012 *Reversibility of Decisions and Retrievability of Radioactive Waste. Considerations for National Geological Disposal Programmes* OECD, NEA No. 7085 https://oecd-nea.org/jcms/pl_14800/reversibility-of-decisions-and-retrievability-of-radioactive-waste

NEA (Nuclear Energy Agency) 2013 *Preservation of Records, Knowledge and Memory Across Generations (RKandM). Literature Survey on Markers and Memory Preservation for Deep Geological Repositories* OECD, NEA/RWM/R(2013)5 http://oecd.org/officialdocuments/publicdisplaydocumentpdf/?cote=NEA/RWM/R(2013)5anddocLanguage=En

NEA (Nuclear Energy Agency) 2020 *Storage of Radioactive Waste and Spent Fuel* OECD, NEA No. 7406 https://oecd-nea.org/jcms/pl_38031/storage-of-radioactive-waste-and-spent-fuel

NWMO (Nuclear Waste Management Organzation) 2004 *Understanding the Choices. The Future Management of Canada's Used Nuclear Fuel* https://nwmo.ca/~/media/Site/Files/PDFs/2015/11/04/17/31/567_DiscussionDocument2.ashx?la=en

NWMO (Nuclear Waste Management Organzation) 2005 *Choosing a Way Forward. The Future Management of Canada's Used Nuclear Fuel. Final Study* https://nwmo.ca/~/media/Site/Files/PDFs/2015/11/04/17/39/2680_nwmo_final_study_nov_2005.ashx

ONR (Office for Nuclear Regulation) 2021 *Low Level Waste Repository* https://onr.org.uk/sites/low-level-waste-repository.htm

OPG (Ontario Power Generation) 2021 *Our Projects. Deep Geologic Repository* https://opg.com/strengthening-the-economy/our-projects/the-deep-geologic-repository/

Rocher M and Železnik N 2020 *Deep Borehole Repository (DBR) of HLW—State of Knowledge* SITEX.Networkreport

Röhlig K-J 2019 The ENTRIA Project (2013–2018). First steps towards sociotechnical radioactive waste management research in Germany *Conflicts, Participation and Acceptability in Nuclear Waste Governance. An International Comparison* vol III ed A Brunnengräber and M R Di Nucci (Wiesbaden: Springer VS) pp 311–22

Röhlig K-J, Chaudry S and Plischke E 2019 *Entsorgungsoptionen für radioaktive Reststoffe: Interdisziplinäre Analysen und Entwicklung von Bewertungsgrundlagen (ENTRIA, 2013–2018): Abschlussbericht. Ergebnisse und Leistungsbilanz (in German, Disposal Options for Radioactive Residues: Interdisciplinary Analyses and Development of Evaluation Principles (ENTRIA, 2013–2018): Final Report. Results and Achievements' Summary)* TU Clausthal http://doi.org/10.21268/20190225-1

SKB (Swedish Nuclear Fuel and Waste Management Co. Svensk Kärnbränslehantering AB) 2021 *The Final Repository SFR* https://skb.com/our-operations/sfr/

STUK (Säteilyturvakeskus. Radiation and Nuclear Safety Authority) 2020 *Disposal of Low and Intermediate-level Waste in Finland* https://stuk.fi/web/en/topics/nuclear-waste/disposal-of-low-and-intermediate-level-waste-in-finland

SÚRAO (Správa úložišť radioaktivních odpadů Radioactive Waste Repository Agency) 2019a *Dukovany Repository* https://surao.cz/en/public/operational-repositories/dukovany-repository/

SÚRAO (Správa úložišť radioaktivních odpadů Radioactive Waste Repository Agency) 2019b *Richard Repository* https://surao.cz/en/public/operational-repositories/richard-repository/

TVO 2018 (Teollisuuden Voima Oyj. Industry Force Corporation) *Operational Waste* https://tvo.fi/en/index/production/nuclearwastemanagement/operationalwaste.html

UK Nirex Ltd 2002 *Description of Long-term Management Options for Radioactive Waste Investigated Internationally* Nirex Report No: N/050

WIPP (Waste Isolation Pilot Plant) 2021 *About Us* https://wipp.energy.gov/about-us.asp

WNA (World Nuclear Association) 2021 *Storage and Disposal of Radioactive Waste (Updated May* 2021) https://world-nuclear.org/information-library/nuclear-fuel-cycle/nuclear-waste/storage-and-disposal-of-radioactive-waste.aspx

Chapter 8

Geologic ('deep') disposal of high-level and other long-lived waste: host rocks, concepts, current international status

Klaus-Jürgen Röhlig

8.1 Geologic ('deep') disposal: basic objectives and requirements

As described in chapter 7 'Disposal and other conceivable strategy 'endpoints' for different types of waste,' 'deep' or 'geologic' disposal is the management option recommended by international organisations especially for high-level waste (HLW), including spent nuclear fuel (SNF) declared as waste. This term means an emplacement of such waste in a geo-engineered system at a depth of at least several hundred meters, with the aim of creating a system whose safety after closure will not depend on supervision or maintenance by humans ('passive safety'). This is not to say that supervision and monitoring should cease, but that the concept is not to rely on it. Measures to allow or to ease retrieval or recovery of the waste might be part of the concept, but the intent is to not retrieve the emplaced material. Generally, the waste to be emplaced should be solid or solidified; approaches to the disposal of liquid waste in deep geologic formations do not satisfy safety and environmental protection objectives. In the following, we will focus on the existing mature concepts for disposal in mined facilities (for disposal in very deep boreholes see, for example, the overview given by Rocher and Železnik 2020).

The overall objectives of disposal are to ensure safety, security, and fairness. IAEA (2018) defines (nuclear) safety as:

> 'The achievement of proper operating conditions, prevention of accidents and mitigation of accident consequences, resulting in protection of workers, the public and the environment from undue radiation risks.'

This definition takes the nuclear perspective, but hazards other than nuclear must be taken into account as well, in particular those arising from chemotoxic substances

doi:10.1088/978-0-7503-3095-4ch8

contained in the waste as well as those related to constructing, operating, and closing a mined repository, including the potential for surface damage.

(Nuclear) security is defined as:

> 'The prevention and detection of, and response to, criminal or intentional unauthorized acts involving nuclear material, other radioactive material, associated facilities or associated activities.'

In a second version of this definition, such acts are specified as 'theft, sabotage, unauthorized access, illegal transfer or other malicious acts' (IAEA 2018).

Compared to other nuclear installations, disposal facilities (repositories) have to meet specific requirements in that safety and security have to be ensured not only during construction, operation (i.e. waste emplacement), and decommissioning (which in their case means closure), but also thereafter: the facility with its inventory and, thus, the hazard potential do not cease to exist after decommissioning. The durations (timeframes) for which safety and security are to be considered are subject to national regulations. There are various motivations for the derivation of such timeframes: one can consider the evolution of the hazard potential over time as well as the human ability (or lack thereof) to make prognoses about the fate of the repository system in the future. The arguments made include a discussion of the evolution of the radiotoxicity of the waste. In the case of surface disposal of short-lived waste, this potential might cease after some hundred years, which lead to concepts that rely on monitoring (oversight) for such timeframes; see chapter 7 'Disposal and other conceivable strategy 'endpoints' for different types of waste'[1].

However, for high-level waste and spent fuel, the radiotoxicity only becomes comparable to that of uranium ore after a couple of hundred thousand years (NEA 2002, 2004, 2009). This seems to coincide with the possible prognoses:

> 'With regard to the requirement of isolation of radioactive waste from the biosphere it can be stated that certain rock formations only show low permeabilities for fluid phases or that they are even water tight in the technical sense due to their physical and chemical properties and to the type of rock formation. Any well-founded predictions of the future evolution of such rock formations and their properties can only be made if the geological setting and its geological history are taken into account. Here, the prediction period is closely related to knowledge of geological evolution in the past. If the evolution of such a geological system can be traced back over many millions of years and can be scientifically interpreted and if furthermore no major

[1] 'For a geological disposal facility, it is possible to provide for safety after closure by means of passive features. It is likewise possible to provide for the safety of a borehole disposal facility after closure by means of passive features, owing to the host geology. In the case of a near-surface disposal facility, actions such as maintenance, monitoring, or surveillance may be necessary for a period of time after closure to ensure safety.' (IAEA 2011)

> changes of the safety-relevant features of this geological system are registered, justified predictions about its future evolution can be made that lie within an order of magnitude of one million years.' (AkEnd 2002)

It could, however, be argued that uranium ore is a hazardous material and that therefore the consideration should go (far) beyond several hundred thousand or one million years. On the other hand, it can be debated for how long responsibilities extend into the (very far) future. Note also that the hazard from highly chemotoxic waste will never cease to exist and that therefore, according to the same type of logic, the safety of its disposal would have to be addressed for an infinite time. A possible conclusion is that:

> 'The time frame of one million years for the evaluation of the long-term safety of high-level waste seems an appropriate compromise between the ethical requirements for long-term responsibility and the limits of practical reason.' (Streffer *et al* 2011)

Debates about the choices of appropriate timeframes are ongoing (and will probably continue in the future). It can, however, be observed that national regulations show some variation concerning this issue, but currently generally address timeframes from several hundred thousand or one million years. In some cases, they ask for certain considerations of the farther future (e.g. about the remaining hazard potential).

Fairness considerations, especially those related to intergenerational fairness, and their implications for issues such as passive safety, accessibility, and retrievability, are briefly discussed in chapter 7 'Disposal and other conceivable strategy 'endpoints' for different types of waste.'

'Disposal options are designed to contain the waste by means of passive engineered and natural features and to isolate it from the accessible biosphere to the extent necessitated by the associated hazard.

[...]

The specific aims of disposal are:

(a) To contain the waste;
(b) To isolate the waste from the accessible biosphere and to reduce substantially the likelihood of, and all possible consequences of, inadvertent human intrusion into the waste;
(c) To inhibit, reduce, and delay the migration of radionuclides at any time from the waste to the accessible biosphere;
(d) To ensure that the amounts of radionuclides reaching the accessible biosphere due to any migration from the disposal facility are such that possible radiological consequences are acceptably low at all times.' (IAEA 2011a).

These aims are valid for all disposal types including deep geologic disposal, the specific feature of which is the objective of durability over long timeframes (see above), to be achieved based on the long-term stability of geologic formations and their favourable features. However, despite being called 'geologic,' such facilities are geo-engineered systems consisting of geologic, geotechnical, and technical components, the latter two often summarised by the term 'engineered barrier system' (EBS). All components have to act in concert: 'Containment and isolation [...] shall be provided by means of a number of physical barriers of the disposal system. The performance of these physical barriers shall be achieved by means of diverse physical and chemical processes together with various operational controls. [...] The overall performance of the disposal system shall not be unduly dependent on a single safety function' (IAEA 2011a).

With regard to the latter point, IAEA (2011a) further specifies: 'Adequate defence in depth has to be ensured by demonstrating that there are multiple safety functions, that the fulfillment of individual safety functions is robust and that the performance of the various physical components of the disposal system and the safety functions they fulfill can be relied upon, as assumed in the safety case and supporting safety assessment.' Examples are given by saying

> 'A safety function may be provided by means of a physical or chemical property or process that contributes to containment and isolation, such as: impermeability to water; limited corrosion, dissolution, leach rate and solubility; retention of radionuclides; and retardation of radionuclide migration.'

Sometimes the term 'multiple barriers' is used, making reference to physical components ('barriers') forming the repository system, such as the waste form, the container, backfill and buffer materials, geotechnical sealing structures, and the geologic formations. In the opinion of the author, however, the term 'multiple safety functions' is more appropriate and illustrative in that it better reflects that properties and processes might evolve over time, and so might their significance and importance in terms of safety. There might even be functions (sometimes called 'reserve functions') which will only become effective if other functions fail or are not as efficient as intended. Such 'reserve functions' can provide safety margins and enhance robustness.

Thus, we understand 'safety function,' to mean 'a property of a component of the disposal system or a process taking place in the disposal system that fulfils the safety-relevant requirements for a safety-related system or subsystem or for an individual component' (EndlSiAnfV 2020). For the time after repository closure, the German Ordinance quoted here defines two pillars of safety: 'safe containment of radioactive waste' within so-called 'essential barriers', and the 'integrity and robustness' of the barriers which should achieve such containment. Note that 'safe containment' here could mean physical containment in the strong sense of the word, or the inhibition, reduction, and delay of radionuclide migration, or a combination of the two. Integrity is defined as the 'preservation of the properties of the barriers of the disposal system relevant for the safe containment of radioactive waste', and robustness is defined as the 'insensitivity of the safety functions of the disposal system and its barriers to internal and external influences and interferences'.

8.2 What constitutes favourable geologic conditions?

Identifying an appropriate site for a deep repository ('siting') is one of the main challenges for radioactive waste management programmes: the concept relies heavily on favourable geologic conditions, which need to be found, and the choice of a site is a societal undertaking. The interplay between the two is further addressed in various chapters of part II, 'Nuclear waste management and society'. In this chapter, we will focus on geologic issues.

When looking for a favourable geology, obviously the pillars of 'safe containment' and 'barrier integrity and robustness' are guiding principles. Moreover, construction, operation (waste emplacement), and closure, i.e. a combination of mining-type and nuclear engineering-type activities must be feasible and safe. Integrity and robustness can best be expected in a regional geologic setting which undergoes, even by geoscientific standards, only very slow changes. Accordingly, the German Site Selection Act (StandAG 2017) defines the following 'exclusion criteria', i.e. conditions under which a region or a site should be excluded from siting:

- Large-scale vertical movements
- Active fault zones
- Safety-relevant impacts from present or past mining activities
- Seismic activity
- Volcanic activity
- Verified presence of young groundwaters

The next consideration when siting a repository concerns the geologic formation into which the waste will be emplaced, the so-called 'host formation' or 'host rock'. Most repository programmes have investigated or are investigating one or more of the following potential host rocks:

- clay or claystones of varying mineralogy and induration (argillaceous media),
- crystalline hard rocks, e.g. plutonites (such as granite) or metamorphic rocks (such as gneiss), and
- rock salt in flat (bedded) or steep (domed) configurations.

Moreover, volcanic tuff is the host rock for the proposed repository at Yucca Mountain (DOE 1998), the future of which is unclear (see below).

Host rock features that are important with respect to the abovementioned requirements are its mechanical properties, its reaction to the heat generated by the emplaced waste, and its hydraulic and hydrogeochemical properties. Some key properties of the three host rock types are summarised in table 8.1, the colour scheme indicates favourable (green), intermediate (yellow), and disadvantageous conditions (red). Of course, the depiction is very cursory: A multitude of features and characteristics are subsumed into a limited number of generalised properties, more detailed information is given e.g. by IAEA (2011b), appendix I. Moreover, the table entries present characteristics which are generally typical for the host rock types, but actual features will vary from site to site. For example, argillaceous media show a considerable variability concerning features with a significant impact on the geomechanical, hydrogeological, and hydrogeochemical situation such as clay content, mineralogy

Table 8.1. Mechanical, thermal, hydraulic, and hydrogeochemical properties of host rock types. The colour scheme indicates favourable (green), intermediate (yellow) and disadvantageous conditions (red). Based on BGR (2007) (modified).

Feature	Rock salt	Argillaceous media	Crystalline (hard) rock
Mechanical strength	Medium	Low to medium	High
Vault stability	No lining needed	Lining needed	Dependent on fissuring / fracturing
In situ stresses	Lithostatically isotropic	Anisotropic	Anisotropic
Deformation behaviour	Viscous (creeping)	Plastic to brittle	Brittle
Heat conductivity	High	Low	Medium
Heat resilience	High	Low	High
Rock solubility	High	Very low	Very low
Permeability	Impermeable	Very low to low	Dependent on fissuring / fracturing
Sorption capacity	Very low	Very high	Medium to high

in general, grain sizes, induration, stratification, and fissuring/fracturing. Particularly the extent and the configuration of fissuring and fracturing are of utmost importance concerning the relevant features of sites with crystalline host formations.

Despite these limitations, the table conveys an important message concerning the choice of host rock, site, and repository concept: there is no such thing as an ideal host rock or site; each has its advantages and shortcomings. Appropriate siting means a balance of the two, and safety concepts and associated technical repository layouts need to take advantage of the former while compensating for the latter.

8.3 Safety concepts and repository layouts for different host rocks

IAEA (2011a) provides a number of guiding principles to be used when developing repository concepts; these concepts should take maximum advantage from the features a repository site has to offer, e.g. with respect to the definition of barriers and safety functions (see above):

> '[...] optimal use has to be made of the safety features offered by the host environment. This has to be done by designing a disposal facility that does not cause unacceptable long-term disturbance of the site, is itself protected by the site and performs safety functions that complement the natural barriers.

The engineered and physical barriers that make up the disposal system are physical entities, such as the waste form, the packaging, the backfill, and the host environment and geological formation. A safety function may be provided by means of a physical or chemical property or process that contributes to containment and isolation, such as:

- impermeability to water;
- limited corrosion, dissolution, leach rate, and solubility; retention of radionuclides;
- and retardation of radionuclide migration.

[...]

The physical elements and their safety functions can be complementary and can work in combination. The performance of a disposal system is thus dependent on different physical elements and on other elements that perform safety functions, which act over different time periods. For example, the roles of the waste package and the host geological formation for a geological disposal facility may vary in different time periods.'

Also, one has to aim at '[...] alternative or additional safety functions that are available if a physical element does not fully perform or another safety function is not fulfilled.'

Note, however, that at least one additional barrier type is particularly important for deep geological repositories: typically, the access paths of these facilities, such as shafts, ramps, and drifts will be closed by geotechnical sealing structures (dams or plugs), which are designed to provide low hydraulic permeability and in some cases geochemical retention. Further, it should be noted that certain backfill materials are sometimes called 'buffers,' since they have to provide e.g. chemical buffering.

The interplay of barriers and safety functions will be defined in a 'safety concept.' According to GRS (2012), a safety concept 'describes, by which geologic features and technical measures the required safety for a repository at a site will be achieved and ensured in the long term.' Based on the safety concept, a site-specific repository layout will be derived in which, *inter alia*, the dimensioning of the disposal mine and the engineered components as well as the material choices and requirements for the latter will be specified.

The terminology used to describe this process in international and national documents varies; the terms used include 'safety strategy/disposal concept' (IAEA 2011b), 'safety strategy/safety concept/system description' (NEA 2013a), and 'safety strategy/system description' (IAEA 2012). Despite the use of different terminologies, what national approaches have in common is that top-down methods are applied in which, starting with general regulation-based principles and objectives, more detailed functions, layouts, and specifications are derived in a stepwise way. In the examples below, the respective national terminology will be used.

Safety after repository closure (post-closure safety, long-term safety) is the ultimate goal of disposal. A prerequisite is, however, that the facility can, and will be, safely constructed, operated, and decommissioned (closed). During all phases of disposal, safety (including radiological protection, mining safety, criticality safety) has to be ensured. Therefore, the safety concept and repository layout have to account for all these aspects.

Various repository layouts are being discussed or are in development (for an overview, see Baldwin *et al* 2008):

- Access to the emplacement level for the operating staff, for the waste to be emplaced, and for ventilation can be provided by shafts, ramps, or a combination of the two.
- The waste container to be emplaced can itself provide shielding from radiation (especially from γ and neutron radiation) for the operating staff by means of an appropriately thick shell consisting of a suitable material. Alternatively, shielding can be provided by the emplacement devices, thus allowing for thinner container shells.
- There are concepts in which the container will be preassembled above ground together with backfill material such as concrete or bentonite, resulting in a so-called 'supercontainer.'
- Waste emplacement might take place from access drifts into vertical or horizontal boreholes or into specially dimensioned vaults or drifts.
- Dependent on all the above, devices for transport from above ground, within the repository mine, and for canister emplacement will be designed. These, in turn, will determine the layout of shafts, ramps, and access drifts (tunnels).

8.3.1 Disposal in crystalline rock: examples from Sweden and Finland (the KBS-3 concept)

KBS-3 (Kärnbränslesäkerhet, nuclear fuel safety) is a concept developed by the Swedish Nuclear Fuel and Waste Management Company SKB for the disposal of spent nuclear fuel (SNF) in crystalline host rocks of the Scandinavian Shield. The advanced SNF disposal projects in Sweden and Finland are using this concept.

The concept takes advantage of the high mechanical strength of crystalline host rocks. Since crystalline rocks contain fractures which might be or might become water conducting and serve as pathways for radionuclides and other contaminants, the concept relies on containment in a container with a design lifetime of more than 100 000 years (King and Lilja 2011). The KBS-3 container, often referred to as the 'copper canister', consists of a nodular cast-iron insert with a 5 cm thick copper shell. It will be surrounded by a backfill made of bentonite (a 'bentonite buffer'), which will provide favourable mechanical, hydraulic, and chemical conditions essential for the integrity of the canister over the time frame of concern. Siting and design should ensure favourable hydraulic, hydrogeochemical, and mechanical conditions favourable for the integrity of the canister-buffer system.

The safety principles, as specified by SKB (2011a, 2011b), are:

- 'The final repository will be located deep down in a long-term stable geological environment to isolate the waste from man and the environment. This reduces the risk that the repository will be impacted by possible societal changes or long-term climatic changes.

- The final repository will be located at a site where the host rock can be assumed to be of little economic interest for future generations, reducing the risk of human intrusion.
- The spent nuclear fuel will be surrounded by multiple safety barriers – engineered and natural.
- The primary safety function of the barriers will be to contain the fuel in the canister.
- If the containment should be breached, the secondary safety function of the barriers will be to retard any release from the repository.
- Engineered barriers will consist of naturally occurring materials that are stable in the long-term in the repository environment.
- The repository will be designed so that the radiation from the spent fuel does not have significant detrimental effects on the properties of the engineered barriers or rock.
- The repository will be designed so that high temperatures, which could ultimately have significant detrimental effects on the properties of the barriers, are avoided.
- The barriers will be passive, i.e. function without human intervention and without active input of materials or energy.'

In both Sweden and Finland, licence applications rely on a reference variant called 'KBS-3V', in which 'V' stands for the emplacement of the canisters in vertical boreholes to be drilled down from the deposition tunnels which will be situated at a depth of several hundred meters. As a potential alternative, the so-called 'KBS-3H' variant is being considered, in which supercontainers with copper canisters surrounded by bentonite buffer and a perforated steel shell will be emplaced in boreholes to be drilled horizontally from a deposition niche.

The construction licence for the Finnish repository at Olkiluoto was granted in 2015; before the start of disposal, a further operational licence is needed, the application process for which is ongoing (STUK 2020a, 2020b). An application for the operational licence was submitted by the operator POSIVA on 30 December 2021 (https://www.posiva.fi/en/index/news/pressreleasesstockexchangereleases/2021/posivasubmitsapplicationforoperatinglicenceforencapsulationandfinaldisposalfacilityofspentnuclearfuel.html [accessed 22 January 2022]). In Sweden, the licensing processes are ongoing. While the Swedish Radiation Safety Authority SSM recommended that the government should grant the licence, the Swedish Lands and Environment Court requested further investigation of several issues concerning the corrosion of the canister's copper shell (Nacka 2018), which has been subject to a long-lasting scientific debate, starting with a publication by Hultquist (1986). In a technical report, the Swedish implementer SKB (2019) addressed the concerns expressed by the Lands and Environment Court. If the licence is granted by the Swedish government, further process permits will have to be issued by the Swedish Radiation Safety Authority, SSM, for construction, trial operations, operation, and closure[2].

[2] After the editorial deadline: In January 2022 the Government decided to grant a licence to construct, possess and operate the repository. SSM has to perform a review process prior to the facility being taken into routine operation. (https://www.stralsakerhetsmyndigheten.se/en/ [accessed 25 March 2022])

Note that the licence applications in both Sweden and Finland are closely linked to the applications for the so-called encapsulation plants, in which the conditioning, i.e. the drying of the spent fuel assemblies (which will come from wet storage), the packaging into the copper canisters and the welding of the canister lids, will take place. In Finland, the encapsulation plant is to be constructed close to the repository site on the Olkiluoto island (Eurajoki municipality); welded canisters can then go underground directly from the encapsulation plant. Olkiluoto is also the site of the nuclear power plant of the same name. In contrast, in Sweden the repository site is going to be at Forsmark (Östhammar municipality), while the encapsulation plant will be constructed in the Oskarshamn municipality which is situated more than 400 km farther south.

Another interesting difference between the two countries is the way in which the two implementers, POSIVA and SKB, utilised underground research laboratories (URLs) in support of their programme. Generally, URLs can serve various purposes, including:

- the development and testing of the technologies used for repository construction, operation, and closure, and
- enhancement of the knowledge about, and the phenomenological understanding of, features and processes relevant to technical feasibility and safety (in particular, post-closure safety).

In addition, they can serve training, communication, and participation purposes. NEA (2013b) broadly distinguishes between two URL types:

- '*Generic URLs.* Facilities that are developed for generic research and testing purposes at a site that *will not be used* for waste disposal, but provide information that may support disposal elsewhere.
- *Site-specific URLs.* Facilities that are developed at a site that *is considered* as a potential site for waste disposal and may, indeed, be a precursor to, or the initial stage of, developing a repository at the site.'

In Finland, a site-specific URL called ONKALO (English 'cavity') was constructed starting in 2004. In addition to the abovementioned typical URL functions, ONKALO served the specific purpose of site characterisation and confirmation (POSIVA 2012a). Currently, the term ONKALO® is used for the whole facility, including the repository under construction. In contrast, the Swedish programme relied on information obtained at the generic URLs at Äspö and Stripa (NEA 2013b). Of course, for every bit of information and detail gathered at generic URLs, the extent of its transferability, applicability, and usability to the site under consideration for actual disposal has to be examined and substantiated.

Hereinafter, the hierarchical top-down approach to safety (see above) will be described in a very exemplary and broad-brush way. POSIVA (2012a) defined a formal requirements management system, *vaatimusten hallintajärjestelmän* (VAHA), with these levels:

- 'Level 1: *stakeholder requirements* that come from laws, decisions-in-principle, regulatory requirements, and other stakeholder requirements;
- Level 2: the *long-term safety principles*, which lead to the definition of the *safety concept* and *safety functions*;
- Level 3: the *performance requirements*, consisting of *performance targets* for the engineered barriers and *target properties* for the host rock, such that the safety functions are fulfilled;

- Level 4: the *design requirements* for the engineered barriers and the underground openings, including *rock suitability classification criteria* (RSC criteria), such that the performance requirements will be met;
- Level 5: the *design specifications*, which are the detailed specifications to be used in design, construction, and manufacture.'

Level 1 requirements include legislation and regulatory guidance. In addition to the top-level principles (implementation in stages, mutually complementary barriers) and criteria such as dose limits, the Guide YVL D.5 (STUK 2013) also becomes more specific by e.g. saying 'that the safety functions provided by the engineered barriers will effectively limit the release of radioactive materials into the bedrock at least for about 10 000 years.' Furthermore, the components of the repository system, such as the waste matrix, the container, its chemical environment, the buffer, the backfill material, and the seals as well as the natural barriers are named and their functions are described broadly and qualitatively.

At level 2, the implementer Posiva (2012a) adapts the regulatory requirements by deriving principles such as the primary containment of the radionuclides by the engineered barriers and the favourable conditions for their performance to be provided by the host rock. This leads to a safety concept with safety functions such as:

- Canister: Ensuring 'a prolonged period of containment of the spent nuclear fuel. This safety function rests first and foremost on the mechanical strength of the canister's cast-iron insert and the corrosion resistance of the copper surrounding it.'
- Buffer:
    - Contributing to predictable and favourable 'mechanical, geochemical, and hydrogeological conditions' for the canister,
    - protecting it 'from external processes', and
    - limiting and retarding 'radionuclide releases in the event of canister failure'
- Deposition tunnel backfill:
    - Contributing to predictable and favourable 'mechanical, geochemical, and hydrogeological conditions for the buffer and the canisters',
    - limiting and retarding 'radionuclide releases in the event of canister failure', and
    - contributing to mechanical stability
- Host rock:
    - Isolating the repository from the surface environment and protecting it,
    - Providing 'favourable and predictable mechanical, geochemical, and hydrogeological conditions for the engineered barriers',
- Repository sealing ('closure')
    - Preventing 'the underground openings from compromising the long-term isolation of the repository,'
    - contributing 'to favourable and predictable geochemical and hydrogeological conditions for the other engineered barriers by preventing the formation of significant water conductive flow paths through the openings,' and
    - limiting and retarding 'inflow to and release of harmful substances from the repository.'

At the further levels, the design basis and specifications are derived based on considerations of 'the different loads and interactions that may act on the repository system at the time of canister emplacement and in the long term,' potential alternative lines of future evolution and conditions, and also on implementation and robustness considerations. While so-called 'performance targets' are derived for the engineered components, these considerations lead to 'target properties' for the host rock. The latter leads to a rock suitability classification (RSC) system which 'is used to identify suitable rock volumes for repository panels and to assess the suitability of deposition tunnels for locating deposition holes and to accept deposition holes for disposal.'

Examples of performance targets for the canister include resistance to corrosion and mechanic loads. For the buffer, the backfills, and the plugs, performance targets concerning their thermal, hydraulic, chemical, and mechanical behaviour are derived. The target properties for the host rock include emplacement depth, restrictions on pH values, chloride concentrations, concentrations of corrosive agents, ionic strengths, salinity, and microbiological conditions as well as the flow regime.

The performance targets then lead to design requirements and specifications for the engineered barriers, e.g. concerning material properties, geometric configuration, and quality control. These are then input to the 'production line' reports that address the actual fabrication of the engineered components (Posiva 2012b).

8.3.2 Disposal in indurated clay: an example from France

In the following, we describe the essentials of a concept developed in France for the disposal of vitrified high-level reprocessing waste and long-lived ILW at the border of the Meuse and Haute-Marne departments in eastern France (Andra 2021). The facility, called Cigéo (Centre Industriel de Stockage Géologique) is planned to be constructed close to the Bure underground research laboratory. The URL served, amongst other purposes, the characterisation of the intended host rock, the Callovo-Oxfordian claystone formation (COX). The disposal site has been chosen in such a way as to ensure the transferability of the data gained at the URL. The application for a construction licence is about to be finalised (Andra 2019) and the application is likely to be submitted in 2021 or 2022 (Hoorelbeke 2021).

The Cigéo concept takes particular advantage of the low permeability and the high sorption capacity of the clay host rock. This means that the host formation plays a central role for the containment of toxic substances and for the delay and attenuation of their migration. Of course, the vaults created during repository construction have to be closed appropriately in order to prevent them from becoming migration pathways. Appropriate siting and design should ensure waste emplacement in sufficiently large and homogeneous clay formations. Specific measures, which will not be addressed in this chapter, are foreseen for ensuring the retrievability of the waste during the operational period (Andra 2016c).

The high-level principles underlying the design are derived from regulatory guidance (ASN 2008, English translation from Andra 2016b):

> 'Protection of human health and the environment constitutes the fundamental safety objective assigned to the disposal of radioactive waste in a deep geological formation. It must be insured against the risks related to the spread of radioactive substances and toxic elements. After closure of the facility, the protection of human health and the environment must not depend on monitoring and institutional checks which cannot be maintained with certainty beyond a limited period.'
>
> Therefore, '[...] the geological medium shall be chosen and the disposal facility shall be designed such that its post-closure safety shall be ensured passively to protect humans and the environment from the radioactive substances and toxic elements contained in the radioactive waste, without the need for intervention.
>
> [...]
>
> In this regard, the concept chosen for the disposal facility should enable the radiological impact to be kept as low as reasonably possible taking account of current scientific knowledge, existing technology, and economic and social factors.'

From this and from the properties of the COX, Andra (2016b) derives, *inter alia*, the following principles:

> 'The disposal system relies particularly on the Callovo-Oxfordian, a natural barrier that plays the main role, and the closure structures of the surface-to-bottom connections, which ensure passive post-closure safety over the long term.
>
> [...]
>
> The global approach to post-closure safety assessment is based on practical expression of the safety functions and associated requirements, analysis of component performance and analysis of the uncertainties related to the scientific and technological knowledge underpinning the design.'
>
> These safety functions are then defined as
>
> - 'isolating the waste from surface phenomena and human actions,' and
> - 'limiting the transfer into the biosphere of the radioactive substances and toxic elements in the waste, controlling the physicochemical degradation of the waste.'

Isolation is to be ensured by the characteristics of the site and the depth of the repository as well as by measures taken to maintain the memory of the facility for as long as possible. *Transfer limitation* relies on 'controlling the physicochemical degradation of the waste, the packages and the engineered components, keeping the radioactive substances and toxic elements as close as possible to their sources, and controlling the transfer pathways that could, in the long term, lead these elements into the biosphere.'

These pathways could be aqueous or gaseous; the former is especially important, because water can not only act as a transport medium for toxic substances, but can also contribute to the degradation of the technical components of the repository, particularly those of the waste form.

In order to limit transfer in the aqueous phase, three further safety functions are defined at a second level, which rely on various characteristics of the site and the host rock:

- 'preventing the circulation of water',
- 'restricting the release of radionuclides and toxic elements and immobilising them in the repository', and
- 'delaying and reducing the migration of radioactive substances and toxic elements released from the packages and then from the disposal cells.'

These three functions are further ensured or supported by more specifically defined functions at a third level:

- 'Preventing the circulation of water' will be achieved by
    - 'limiting water flows, distinguishing between the flow of water from the Callovo-Oxfordian host formation drained by the underground facility, and the flow of water circulating via the surface-to-bottom connections between the disposal facility and the overlying aquifer formations [...];
    - limiting the speed of circulation of this water from the cells to the top of the Callovo-Oxfordian through the surface-to-bottom connections.'
- Sub-functions supporting the function of 'restricting the release of radionuclides and toxic elements and immobilising them in the repository' are
    - 'protecting the waste from water;
    - restricting the dissolution of the radionuclides and toxic chemicals;
    - restricting the mobility of the radionuclides and toxic chemicals.'
- 'Delaying and reducing the migration of radioactive substances and toxic elements' will be achieved by
    - retarding and mitigating radionuclide migration along repository drifts as well as,
    - in the geologic host formation.
    - In addition, the potential for dilution and dispersion in surrounding formations has to be preserved.

The repository is designed to take maximum advantage of the host rock properties, as well as to control the risk of self-sustaining nuclear chain reactions (criticality) and to ensure safe operational conditions. The principles of the latter are derived from regulations related to various areas, e.g. to road tunnels, mining installations, and nuclear installations, as well as to several industrial standards (Andra 2016a). They concern issues such as fire risks, workers' safety, and radiation protection. The safety principles to be adhered to include:

- 'contain radioactive substances to protect against the risk of their dispersion,'
- 'protect people from exposure to ionising radiation,'

- 'manage safety with regard to the criticality risk,'
- 'remove the heat produced by waste,' and
- 'remove gases formed by radiolysis in order to manage explosion risks.'

Andra (2016a) emphasizes the need to consistently address issues that affect repository operation and long-term safety, and the importance of the integration of organisational and human factors (OHF) to support both.

For each of the safety functions mentioned above, Andra (2016a, 2016b) defines supporting 'general principles in terms of choice of site and design.' These relate to the general location of the repository as well as the repository layout, and the choice of materials to be used. The general location is chosen to take advantage of the lateral and vertical extension of the COX formation, its amenability to characterisation, its long-term stability, and its hydrogeological and geochemical characteristics. The properties named include homogeneity, low hydraulic permeability, low hydraulic gradients, stagnant water, low diffusion coefficients, low radionuclide solubilities, and high sorption capacities as well as favourable geomechanical properties and the self-sealing capacity of the host rock.

The specifications for the repository design derived from the abovementioned principles extend to:

- the waste packages, i.e. the waste containers and the inventory itself, and
- the repository architecture, i.e. the vault arrangement and geometry, the linings, backfills, and seals.

Hereinafter, only a few examples are given:

- The general layout is chosen to avoid creating driving forces for water flow, to take maximum advantage of the COX layer thickness in order to take advantage of its containment capacity.
- Disposal cells, connecting drifts and disposal sections will be oriented in a way that the seals can be emplaced accounting for the major horizontal stress in order to ensure maximum seal performance and to minimise damage to the host rock.
- Heat-generating waste has to be emplaced at a defined distance from other waste types.
- Appropriate materials are chosen for waste packages, linings, backfills, and seals in order to provide a favourable geochemical environment. For the latter, a high proportion of clay materials will be used.
- HLW containers in particular (low-carbon unalloyed steel) have to remain leak-tight for the initial phase after emplacement, during which temperatures are high.
- For ILW packages, low-solubility salts will be used in order to support immobilisation.

8.3.3 Disposal in rock salt: an example from Germany

Rock salt is one of the host rock formations considered for HLW and SNF disposal in Germany. In the past, the Gorleben salt dome in particular was considered.

Gorleben was subject to decades of controversy. It is no longer considered a potential site in the ongoing siting procedure in Germany. However, many historical research and development results related to disposal in rock salt were based on data from investigations of the site by various methods. In particular, the investigation mine that operated there in the past, which can be considered as an attempt to establish a site-specific URL, is worth mentioning. From 2010 to 2013, a Preliminary Safety Assessment was carried out (GRS 2013a), in which a safety concept for disposal in rock salt (steep formation) was developed (GRS 2012, 2013b). The concept was later further developed (DBE TECHNOLOGY 2017a) and served as a conceptual basis for generic safety concepts for disposal in other host rocks such as bedded salt, claystone, and crystalline rock (BGE TECHNOLOGY 2018, DBE TECHNOLOGY 2017b, 2016).

> The concept takes particular advantage of the fact that undisturbed rock salt is practically impermeable. Moreover, the fact that rock salt creeps under the influence of rock pressure results in so-called 'convergence,' i.e. the closing of vaults over time. The latter is especially important for the concept of geotechnical barriers, particularly the use of crushed salt for backfill. The idea is that the crushed salt will become compacted under pressure and will develop features increasingly similar to those of undisturbed rock salt. Until the backfill is sufficiently compacted, drift and shaft seals have to prevent liquids from getting access to the waste. Appropriate siting and design have to ensure that heterogeneous salt formations and formations with considerable brine inclusions are avoided and that salt dissolution by groundwater ('subrosion') is sufficiently limited. In accordance with the regulations (BMU 2010), specific measures, which will not be addressed in this chapter, were foreseen that ensure the retrievability of the waste during the operational period and ease recovery for a time frame of 500 years.

The Gorleben safety concept is described top-down, starting with (regulatory) requirements, on the basis of which governing principles are derived. These lead to basic requirements, then to target definitions, and finally to concrete (design) measures. At the time the concept was developed, the predecessor of the current valid Safety Requirements (EndlSiAnfV 2020) was in place (BMU 2010), but the requirements that are of central importance for this chapter remained basically unchanged. They include:

- The waste has to be concentrated and contained in a so-called 'containment-providing rock zone' (German: 'einschlusswirksamer Gebirgsbereich' (ewG))[3].

[3] The ewG is defined as the 'part of the repository system which, in conjunction with the technical seals (shaft seals, cavern sealing structures, dam structures, backfill, …), ensures containment of the waste.' (BMU 2010). The unofficial translation (BMU 2010) uses the term 'isolation rock zone.' In this text, the technically more appropriate term 'containment-providing rock zone' is used. The current valid safety requirements (EndlSiAnfV 2020) require containment within so-called 'essential barriers' which could either be an ewG or, for crystalline host rocks, a system of technical and geotechnical barriers.

- Releases from the repository must, at a maximum, lead to negligible increases of radiation exposure compared to natural conditions.
- This has to be ensured without relying on human intervention or maintenance.
- A robust barrier system has to be created (robustness meaning insensitivity to internal and external influences and interference).
- The integrity of the 'containment-providing rock zone,' i.e. the preservation of its containment capabilities, has to be ensured over the assessment time frame of one million years.
- The disturbance of the 'containment-providing rock zone' by the repository vaults has to be kept to a minimum.
- Uncontrolled chain reactions (criticality) have to be prevented.

The governing principles derived from these requirements are:

- As far as possible, the waste has to be contained in a 'containment-providing rock zone.'
- Containment should become effective immediately after repository closure and has to be ensured over the time frame of concern without human intervention or maintenance.
- The containment has to be ensured preferably by avoiding or at least strongly limiting the access of liquids to the waste.

While the first two principles are directly derived from the (generic) safety requirements, the third is specific for rock salt as a host rock and its specific properties (see above). The basic requirements laid out in the concept combine the generic safety requirements and the ideas specific to the host rock:

- Containment: the waste should be contained as quickly as possible by the host rock in combination with the geotechnical barriers.
- Integrity/independence from post-operative care: the 'containment-providing rock zone' will remain existent and its barrier function will not be impaired by internal or external processes.
- Criticality safety: criticality has to be prevented.

The containment requirement above leads to the definition of the following targets:

1. For the likely system evolutions, no liquid or only very limited amounts of liquid should come into contact with the waste.
2. For less likely evolutions, only limited amounts of liquid should come into contact with the waste.
3. If any toxic substances are mobilised, their migration in both the liquid and gaseous phases should be hindered by chemical and physical processes.
4. The properties of the host rock and the technical components important for containment should be amenable to appropriate good characterisation.
5. No corrective action should become necessary after repository closure.
6. The technical components should be designed in such a way that their safety functions will be ensured despite corrosion and degradation processes, ageing processes, and external loads (robustness).

7. The immediate and durable containment should be ensured by a staggered system of barriers which act complementarily over time and, in part, in a diverse and redundant way.
8. The repository should be divided into sections that can be sequentially loaded with waste, backfilled, and closed off from the other parts of the mine. Interactions between different waste types which might impair safety functions should be minimized as far as possible.

These requirements lead to a set of 11 measures, including the geometrical layout of the repository in relation to the characteristics of the site, the use of appropriate mining techniques, quality-assured properties of seals that rely on diverse mechanisms, and an appropriate strategy for waste emplacement and backfilling. A further five targets are derived from the requirement for integrity/independence from post-operative care, which in turn lead to five further measures which again relate to the geometric layout, but also to a temperature limitation. Similarly, a target concerning criticality safety is defined, which leads to measures concerning canister loading with waste and neutron absorbers and measures concerning the geometric configuration of the repository.

8.3.4 Summary and overview

The overall objective of deep (geologic) disposal is the protection of the inhabited environment from the risks caused by toxic, particularly radiotoxic, substances contained in the waste by passive means, i.e. without a dependence on human actions such as monitoring or maintenance, after repository closure. This is not to say that e.g. monitoring should not take place, but just that the system should not rely on it. Broadly, the protection will rely on isolation and containment of the waste. If complete containment cannot be achieved, any migration of the radionuclides should inhibited, reduced, and delayed.

Disposal should take place in regional geologic settings in which long-term stability is to be expected; potentially disturbing effects, such as tectonic movements, earthquake hazards, uplift, and volcanism are to be avoided. The properties of the host rock, i.e. the formation into which the waste will be emplaced, are of particular importance: they are crucial for achieving the objectives mentioned above. A safety concept has to be derived, which specifies which host rock properties can be taken advantage of, which are to be compensated for, and how this can be achieved by developing an appropriate repository layout and choosing the appropriate technical and geotechnical barriers. The terminology and the specifics of the safety concept vary from country to country, but generally the approach is to derive requirements and specifications for repository components from more general principles in a top-down way. In the previous sections, three prototypes for such concepts were presented for different host rocks, with a focus on measures that achieve long-term (post-closure) safety:

In the KBS-3 concept for disposal in *crystalline rock*, the *containers* have to ensure *containment* over about 100 000 years or more. Mechanical, hydraulic, and chemical *stability* (and thus barrier integrity) is to be provided by the bentonite buffer to be emplaced in a stable geology. In the case of failure of the container/buffer system, *migration retardation* is to be ensured by the bentonite buffer, the seals, and the geologic environment.

In the French concept of disposal in *indurated clay*, the containers have to ensure containment for a much shorter time (about 1000 years). Then, based on the lack of groundwater movement, *migration retardation* is to be ensured by the *host rock* and the geotechnical barriers, the former of which is also crucial for the *isolation* and the *stability* of the system.

In the German concept for *rock salt*, *containment* is to be achieved by the *host rock* in conjunction with geotechnical barriers. The geology has to ensure *isolation* and *stability*.

The requirements related to retrievability (and the degree to which it has to be demonstrated) vary from country to country. They also have to be fulfilled by appropriate designs without impairing the fulfilment of the abovementioned safety objectives.

8.4 Deep geologic disposal of HLW and SNF: current international status

Amongst the more than 30 countries currently using nuclear power for electricity production (WNA 2020), there are only some with disposal programmes for HLW and/or SNF worth mentioning:

As explained above in sections 8.3.1 and 8.3.2, **Finland**, **Sweden**, and **France** have advanced programmes: All three have ongoing licensing procedures for repositories at designated sites. Finland (at the Olkiluoto site) and Sweden (at the Forsmark site) will dispose of SNF in crystalline rock, while France plans to construct a repository for vitrified HLW from reprocessing and long-lived ILW in indurated clay (at the Bure site). In the case of Finland, repository construction is ongoing. All three envisage starting operation within this decade[4].

Sites for disposal in crystalline rock were identified in the **Russian Federation** (at the Yeniseiskiy site in the Krasnoyarsk Krai, Kochkin 2019) and in the **People's Republic of China** (at the Xinchang site in the Beishan area, Wang *et al* 2018). In both cases, site-specific URLs are planned and, in the case of China, the URL construction work has begun (China Daily 2021). Implementation might take place in the 2050s.

In **Canada**, **Switzerland**, and the **Czech Republic**, site selection processes are underway:

In Canada, two potential sites were identified for SNF disposal: Ignace in northern Ontario (crystalline rock) and South Bruce in southern Ontario (argillaceous limestone). Operation is planned to begin in the 2040s (NWMO 2021).

In Switzerland, three potential regions for HLW and SNF disposal in Northern Switzerland (host rock Opalinus clay) were identified (Jura Ost, Nördlich Lägern,

[4] After the editorial deadline: On 30 December 2021, the Finnish implementer POSIVA applied for a operating license for the encapsulation and final disposal facility, cf. https://www.posiva.fi/en/index/news/pressreleasesstockexchangereleases/2021/posivasubmitsapplicationforoperatinglicenceforencapsulationandfinaldisposalfacilityofspentnuclearfuel.html. On 27 January 2022, the Swedish government approved the final disposal system and siting, cf. https://www.skb.com/news/the-government-approves-skbs-final-repository-system/ [both websites accessed 27 January 2022].

and Zürich Nordost). All of them can also host a LILW repository; it is yet to be decided whether or not HLW and LILW will be disposed of in two separate repositories at different sites or in a combined repository at one site. SNF/HLW disposal is planned to commence in the 2060s (Nagra 2018, SFOE 2021).

The Czech Republic has identified four potential sites for SNF disposal in crystalline rock (Březový potok, Horka, Hrádek, and Janoch, SÚRAO 2020). Operational commencement is planned for the 2060s.

Germany, Japan, and the **United Kingdom** are at the beginning of their siting procedures. Germany plans to identify a site for HLW/SNF disposal in 2031, which might result in the operational commencement in the 2050s. Rock salt, claystone, and crystalline rocks are considered in the siting procedure (StandAG 2017). The United Kingdom has started a consent-based siting process with the vision of starting operation in the 2040s. No host rock type has been determined so far (RWM 2020). **Japan** initiated a multi-stage site selection procedure in 2002 and received applications for further literature surveys from two municipalities in 2020 (NUMO 2020a, 2020b, 2020c).

In the **United States**, SNF disposal at the Yucca Mountain site (Nevada) has been considered a possibility in the past. A license was applied for in 2008. During the Obama administration, the work on Yucca Mountain was halted and a Commission was assigned and tasked with the development of a perspective for the nuclear waste management programme. In its report, the commission provided advice on such a programme without taking a position concerning Yucca Mountain or any other potential site (BRC 2012). During the Trump administration, the momentum of the program's development was lost. It remains to be seen how the Biden administration will address the issue (NWTRB 2021).

Belgium is investigating disposal in the Boom plastic clay formation at the Hades URL; the Ypresian clays are a potential alternative (Smith *et al* 2009). However, so far there has been no official decision confirming that geologic disposal will be the end point of the waste management strategy. The implementer ONDRAF.NIRAS made a proposal for geologic disposal without naming a potential site or region (ONDRAF 2018). No definitive decision on a management end point for SNF and HLW has been made so far (ONDRAF 2020).

In the **Netherlands**, disposal in the Boom clay formation is being investigated; disposal in bedded salt is a potential alternative. No siting decision will be taken in the near future (COVRA 2018).

Other countries with less advanced geologic disposal programmes for their SNF/HLW but ongoing research and development are **Hungary**, the **Republic of China** (Taiwan), and **South Korea** (NEA 2020, IAEA 2020).

Bibliography

AkEnd (Arbeitskreis Auswahlverfahren Endlagerstandorte. Committee on a Site Selection Procedure for Repository Sites) 2002 *Site Selection Procedure for Repository Sites. Recommendations of the AkEnd —Committee on a Site Selection Procedure for Repository Sites* http://cienciaensocietat.org/upimages/File/Deliberativa_2/10-Site%20Selection%20Procedure%20for%20Repository%20Sites.pdf

Andra (Agence nationale pour la gestion des déchets radioactifs) 2016a *Safety Options Report—Operating Part (DOS-Expl)* CG-TE-D-NTE-AMOA-SR1-0000-15–0060 https://international.andra.fr/sites/international/files/2019-03/Safety%20Options%20Report%20-%20Operations_2.pdf

Andra (Agence nationale pour la gestion des déchets radioactifs) 2016b *Safety Options Report—Post-Closure Part (DOS-AF)* CG-TE-D-NTE-AMOA-SR2-0000-15–0062 https://international.andra.fr/sites/international/files/2019-03/Safety%20Options%20Report%20-%20Post-Closure_1.pdf

Andra (Agence nationale pour la gestion des déchets radioactifs) 2016c *Dossier D'options Techniques de Récupérabilité (DORec)* CG-TE-D-NTE-AMOA-RV0-0000-15–0059/A https://andra.fr/sites/default/files/2018-11/CG-TE-D-NTE-AMOA-RV0-0000-15–0059-A%20DOREC.pdf

Andra (Agence nationale pour la gestion des déchets radioactifs) 2019 *Activity Report 2019* https://international.andra.fr/sites/international/files/2020-08/20200616-AndraRA2019-maquette-PAP-HD_EN_BD.pdf

Andra (Agence nationale pour la gestion des déchets radioactifs) 2021 *Cigeo* https://international.andra.fr/solutions-long-lived-waste/cigeo

ASN (Autorité de sûreté nucléaire) 2008 https://asn.fr/Professionnels/Les-Guides-de-l-ASN/Guide-de-l-ASN-n-1-Stockage-definitif-des-dechets-radioactifs-en-formation-geologique-profonde

Baldwin T, Chapman N and Neall F 2008 Geological disposal options for high-level waste and spent fuel *Report for the UK Nuclear Decommissioning Authority* https://rwm.nda.gov.uk/publication/geological-disposal-options-for-high-level-waste-and-spent-fuel-january-2008/

BGE TECHNOLOGY 2018 KOSINA *Concept Developments for a Generic Repository for Heat-generating Waste in Bedded Salt Formations in Germany* BGE TEC 2018-13 https://bge-technology.de/fileadmin/user_upload/MEDIATHEK/f_e_berichte/KOSINA-Project_Concept-developments-for-a-generic-repository-for-heat-generating-waste-in-bedded-salt-formations.pdf

BRC (Blue Ribbon Commission on America's Future) 2012 *Report to the Secretary of Energy* https://energy.gov/sites/prod/files/2013/04/f0/brc_finalreport_jan2012.pdf

BGR (Bundesanstalt für Geowissenschaften und Rohstoffe) 2007 *Endlagerung radioaktiver Abfälle in Deutschland. Untersuchung und Bewertung von Regionen mit potenziell geeigneten Wirtsgesteinsformationen (in German, Federal Institute for Geosciences and Natural Resources: Final Disposal of Radioactive Wastes in Germany. Investigation and Evaluation of Regions with Potentially Suitable Host Rock Formations)* https://bgr.bund.de/DE/Themen/Endlagerung/Downloads/Charakterisierung_Wirtsgesteine_geotech_Barrieren/4_Wirtsgesteinsuebergreifend/2007-04-00_BGR_wirtsgest_dtl.html?nn=9999498

BMU (Bundesministerium für Umwelt, Naturschutz und Reaktorsicherheit) 2010 *Safety Requirements Governing the Final Disposal of Heat-generating Radioactive Waste: As at 30 September 2010* https://bmu.de/fileadmin/bmu-import/files/english/pdf/application/pdf/sicherheitsanforderungen_endlagerung_en_bf.pdf

China Daily 2021 *Construction of Radioactive Waste Disposal Lab Underway* https://chinadaily.com.cn/a/202104/08/WS606e3b7ba31024ad0bab4108.html

COVRA (Centrale Organisatie Voor Radioactief Afval) 2018 https://covra.nl/app/uploads/2019/08/Opera-SafetyCase-samenvatting.pdf

DBE TECHNOLOGY 2016 *Projekt CHRISTA. Machbarkeitsuntersuchung zur Entwicklung einer Sicherheits- und Nachweismethodik für ein Endlager für Wärme entwickelnde radioaktive*

Abfälle im Kristallingestein in Deutschland (in German, Project CHRISTA. Feasibility Study for the Development of a Safety and Demonstration Methodology for a Repository for Heat-generating Radioactive Wastes in Germany) TEC-20–2016-AB https://bge-technology.de/fileadmin/user_upload/MEDIATHEK/f_e_berichte/CHRISTA-Projekt_Abschlussbericht.pdf

DBE TECHNOLOGY 2017a *Evaluation of Methods and Tools to Develop Safety Concepts and to Demonstrate Safety for an HLW Repository in Salt. Final Report* TEC-03–2017-AB https://bge-technology.de/fileadmin/user_upload/MEDIATHEK/f_e_berichte/KOMTESSA-Project_Evaluation_of_methods_and_tools_to_develop_safety_concepts_in_salt.pdf

DBE TECHNOLOGY 2017b *Projekt ANSICHT. Sicherheits-und Nachweismethodik für ein Endlager im Tongestein in Deutschland. Synthesebericht (in German, Project ANSICHT. Safety and Demonstration Methodology for a Repostory in Clay Rock in Germany. Synthesis Report)* TEC-19–2016-AB https://bge-technology.de/fileadmin/user_upload/MEDIATHEK/f_e_berichte/Ansicht/01_ANSICHT_Projekt-Synthesebericht.pdf

DOE (Department of Energy) 1998 *Viability Assessment of a Repository at Yucca Mountain* https://energy.gov/downloads/viability-assessment-repository-yucca-mountain-1

EndlSiAnfV (Endlagersicherheitsanforderungsverordnung) 2020 *Verordnung über Sicherheitsanforderungen an die Endlagerung hochradioaktiver Abfälle (in German, Ordinance on Safety Requirements for the Disposal of High-Level Radioactive Waste)* http://gesetze-im-internet.de/endlsianfv/ (quotations from the non-binding translation by the German Federal Ministry for Environment, Nature Conservation and Nuclear Safety BMU)

GRS (Gesellschaft für Anlagen- und Reaktorsicherheit gGmbH) 2012 *Sicherheits- und Nachweiskonzept. Bericht zum Arbeitspaket 4. Vorläufige Sicherheitsanalyse für den Standort Gorleben (in German, Safety and Demonstration Concept. Report for Working Package 4. Preliminary Safety Assessment for the Gorleben Site)* GRS-277 https://grs.de/sites/default/files/pdf/GRS-277.pdf

GRS (Gesellschaft für Anlagen- und Reaktorsicherheit gGmbH) 2013a *Gorleben—Preliminary Safety Assessment for the Gorleben Site (VSG)* https://grs.de/en/content/gorleben

GRS (Gesellschaft für Anlagen- und Reaktorsicherheit gGmbH) 2013b *Synthesebericht für die VSG. Bericht zum Arbeitspaket 13. Vorläufige Sicherheitsanalyse für den Standort Gorleben (in German, Synthesis VSG report. Report for Working Package 13. Preliminary Safety Assessment for the Gorleben Site)* GRS-290 https://grs.de/sites/default/files/pdf/GRS-290.pdf

Hoorelbeke J-M 2021 Personal communication

Hultquist G 1986 Hydrogen evolution in corrosion of copper in pure water *Corros. Sci.* **26** 173–7

IAEA (International Atomic Energy Agency) 2011a *Disposal of Radioactive Waste Specific Safety Requirements* IAEA Safety Standards Series No. SSR-5 https://www-pub.iaea.org/MTCD/Publications/PDF/Pub1449_web.pdf

IAEA (International Atomic Energy Agency) 2011b *Geological Disposal Facilities for Radioactive Waste* Specific Safety Guide, IAEA Safety Standards Series No. SSG-14 https://www-pub.iaea.org/MTCD/Publications/PDF/Pub1483_web.pdf

IAEA (International Atomic Energy Agency) 2012 *The Safety Case and Safety Assessment for the Disposal of Radioactive Waste* Specific Safety Guide, IAEA Safety Standards Series No. SSG-23 https://www-pub.iaea.org/MTCD/Publications/PDF/Pub1553_web.pdf

IAEA (International Atomic Energy Agency) 2018 *IAEA Safety Glossary. Terminology used in Nuclear Safety and Radiation Protection. 2018 edn* https://www-pub.iaea.org/MTCD/Publications/PDF/PUB1830_web.pdf

IAEA (International Atomic Energy Agency) 2020 *The IAEA Online Information Resource for Radioactive Waste Management* https://newmdb.iaea.org/

King F and Lilja C 2011 Scientific basis for corrosion of copper in water and implications for canister lifetimes *Corros. Eng. Sci. Technol.* **2011** 153–8

Kochkin B T 2019 Investigating the geological environment at the Yeniseisky site: tasks for the current stage of the disposal project *Radioactive Waste* **2** 76–91 http://eng.radwaste-journal.ru/docs/journals/7/investigatingitheigeologicalienvironmentiatitheiyeniseiskyisite.pdf

Nacka T 2018 Mark-och miljödomstolen Mark-och miljödomstolens yttrande, Mål nr M 1333-11 (in Swedish, Nacka District Court, Land and Environmental Court, *Opinion of the Environmental and Environmental Court, Case no. M 1333-11* https://www.mkg.se/uploads/Aktbilagor/842_Nacka_TR_M_1333-11_MMDs_yttrande_till_regeringen_180123.pdf, unofficial translation of the summary under https://www.mkg.se/uploads/Summary_opinion_Swedish_Environmental_Court_regarding_proposed_final_repository_spent_nuclear_fuel_Forsmark_Jan_23_2018_(unofficial_translation_MKG).pdf

Nagra (Nationale Genossenschaft für die Lagerung radioaktiver Abfälle. National Cooperative for the Disposal of Radioactive Waste) 2018 *Siting Regions* https://nagra.ch/en/sitingregions.htm

NUMO (Nuclear Waste Management Organization) 2020a *The Repository Site Selection Process* https://numo.or.jp/en/jigyou/new_eng_tab03.html

NUMO (Nuclear Waste Management Organization) 2020b *Acceptance by Kamoenai Village of Proposal for Literature Survey* https://numo.or.jp/en/what/topics_201009_2.html

NUMO (Nuclear Waste Management Organization) 2020c *Receipt of Application for Literature Survey from Suttu Town in Hokkaido* https://numo.or.jp/en/what/topics_201009_1.html

NWTRB (U.S. Nuclear Waste Technical Review Board) 2021 *Six Overarching Recommendations for How to Move the Nation's Nuclear Waste Management Program Forward* A Report to the U.S. Congress and the Secretary of Energy https://nwtrb.gov/docs/default-source/reports/six-recommendations-report_508c.pdf?sfvrsn=18

NEA (Nuclear Energy Agency) 2002 The handling of timescales in assessing post-closure safety of deep geological repositories *Workshop Proceedings (Paris, France, 16–18 April 2002)* (Paris: OECD Publishing)

NEA (Nuclear Energy Agency) 2004 *The Handling of Timescales in Assessing Post-closure Safety Lessons Learnt from the April 2002 Workshop in Paris, France* OECD, NEA No. 4435 https://oecd-nea.org/jcms/pl_13798/the-handling-of-timescales-in-assessing-post-closure-safety

NEA (Nuclear Energy Agency) 2009 *Considering Timescales in the Post-closure Safety of Geological Disposal of Radioactive Waste* NEA OECD No. 6424 https://oecd-nea.org/jcms/pl_14446

NEA (Nuclear Energy Agency) 2013a *The Nature and Purpose of the Post-closure Safety Cases for Geological Repositories* OECD, NEA/RWM/R(2013)1 https://oecd-nea.org/jcms/pl_19258

NEA (Nuclear Energy Agency) 2013b *Underground Research Laboratories (URL)* OECD, NEA No. 7532 https://oecd-nea.org/rwm/reports/2013/78122-rwm-url-brochure.pdf

NEA (Nuclear Energy Agency) 2020 *Management and Disposal of High-level Radioactive Waste: Global Progress and Solutions* OECD, NEA/RWM/R(2013)2 https://oecd-nea.org/jcms/pl_32567/management-and-disposal-of-high-level-radioactive-waste-global-progress-and-solutions

NWMO (Nuclear Waste Management Organization) 2021 *Implementing Adaptive Phased Management 2021 to 2025* https://nwmo.ca/~/media/Site/Reports/2021/03/15/21/05/Implementing-Adaptive-Phased-Management-2021-to-2025.ashx?la=en

ONDRAF (Organisme national des déchets radioactifs et des matières fissiles enrichies) 2018 *Projet de Plan. Avant-projet d'arrêté royal établissant le processus d'adoption de la politique nationale relative à la gestion à long terme des déchets radioactifs conditionnés de haute activité et/ou de longue durée de vie et définissant la solution de gestion à long terme de ces déchets* https://ondraf.be/sites/default/files/2020-04/Projet%20de%20Plan_FR_def.pdf

ONDRAF (Organisme national des déchets radioactifs et des matières fissiles enrichies) 2020 *Magazine ONDRAF Décembre 2020 Numéro 12* https://ondraf.be/sites/default/files/2020-12/Magazine%20ONDRAF%20d%C3%A9cembre%202020_def.pdf

Posiva O 2012a *Safety Case for the Disposal of Spent Nuclear Fuel at Olkiluoto—Synthesis 2012* POSIVA 2012-2 https://posiva.fi/en/index/media/reports.html

Posiva O 2012b *Safety Case for the Disposal of Spent Nuclear Fuel at Olkiluoto—Design Basis 2012* POSIVA 2012-3 https://posiva.fi/en/index/media/reports.html

Rocher M and Železnik N 2020 (eds) *Deep Borehole Repository (DBR) of HLW—State of knowledge* SITEX.Networkreport

RWM (Radioactive Waste Management Ltd) 2020 *Guidance Geological Disposal. A Permanent Solution for the UK's Higher-activity Radioactive Waste* https://gov.uk/guidance/geological-disposal

Smith P, Cornélis B, Capouet M and Van Geet M 2009 *The Long-term Safety Strategy for the Geological Disposal of Radioactive Waste* SFC1 Level 4 Report: Second Full Draft https://researchgate.net/publication/331674249_The_Long-Term_Safety_Strategy_for_the_Geological_Disposal_of_Radioactive_Waste_SFC1_level_4_report_second_full_draft

SFOE (Swiss Federal Office of Energy) 2021 *Deep Geological Repositories Sectoral Plan* https://bfe.admin.ch/bfe/en/home/supply/nuclear-energy/radioactive-waste/deep-geological-repositories-sectoral-plan.html

SKB (Svensk Kärnbränslehantering AB. Swedish Nuclear Fuel and Waste Management Company) 2011a *Application for Licence Under the Nuclear Activities Act for Construction, Ownership and Operation of a Nuclear Facility for the Final Disposal of Spent Nuclear Fuel and Nuclear Waste* http://skb.com/wp-content/uploads/2017/01/1282973-KTL-ans%C3%B6kan-p%C3%A5-engelska.pdf

SKB (Svensk Kärnbränslehantering AB. Swedish Nuclear Fuel and Waste Management Company) 2011b *Application for Permit Under the Environmental Code (1998:808) for Facilities in an Integral System for Final Disposal of Spent Nuclear Fuel and Nuclear Waste* http://skb.com/wp-content/uploads/2017/01/1282835-MB-ans%C3%B6kan-p%C3%A5-engelska.pdf

SKB (Svensk Kärnbränslehantering AB. Swedish Nuclear Fuel and Waste Management Company) 2019 *Supplementary Information on Canister Integrity Issues* SKB Technical Report TR-19-15 https://skb.com/wp-content/uploads/2021/02/TR-19-15.pdf

StandAG (Standortauswahlgesetz. Site Selection Act) 2017 *Gesetz zur Suche und Auswahl eines Standortes für ein Endlager für hochradioaktive Abfälle (in German, Act for the Search for, and the Selection of, a Site for a Repository for Highly Radioactive Wastes)* https://gesetze-im-internet.de/standag_2017/

Streffer C, Gethmann C F, Kamp G, Kröger W, Rehbinder E, Renn O and Röhlig K-J 2011 *Radioactive Waste. Ethics of Science and Technology Assessment* (Berlin: Springer)

SSM (Strålsäkerhetsmyndigheten. Swedish Radiation Safety Authority) 2018 *Pronouncement on Licence Applications for Permission to Develop Facilities for Final Management of Spent Nuclear Fuel* SSM2011-1135-23 https://stralsakerhetsmyndigheten.se/contentassets/078506f952ae4357847628edcc1785a4/pronouncement-on-licence-applications-for-permission-to-develop-facilities-for-final-management-of-spent-nuclear-fuel-statement-of-the-swedish-radiation-safety-authority

STUK (Säteilyturvakeskus. Radiation and Nuclear Safety Authority) 2013 *Guide YVL D.5, Disposal of Nuclear Waste* https://ohjeisto.stuk.fi/YVLold/YVLD.5en_2013-11-15.pdf

STUK (Säteilyturvakeskus. Radiation and Nuclear Safety Authority) 2020a Regulatory oversight of nuclear safety in Finland *Annual Report 2019* https://julkari.fi/bitstream/handle/10024/139827/stuk-b248.pdf?sequence=1andisAllowed=y

STUK (Säteilyturvakeskus. Radiation and Nuclear Safety Authority) 2020b Joint convention on the safety of spent fuel management and on the safety of radioactive waste management *7th Finnish National Report as referred to in Article 32 of the Convention* https://julkari.fi/bitstream/handle/10024/140601/stuk-b259.pdf?sequence=1andisAllowed=y

SÚRAO (Správa úložišť radioaktivních odpadů. Radioactive Waste Repository Agency) 2020 *Site Shortlisting. Deep Geological Repository Site Shortlisting Information Page* https://surao.cz/en/site-shortlisting/

Wang J, Chen L, Su R and Zhao X 2018 The Beishan underground research laboratory for geological disposal of high-level radioactive waste in China: planning, site selection, site characterization and *in situ* tests *J. Rock Mechan. Geotech. Eng.* **10** 411–35

WNA (World Nuclear Association) 2020 *Nuclear Share Figures, 2009–2019* https://world-nuclear.org/information-library/facts-and-figures/nuclear-generation-by-country.aspx

IOP Publishing

Nuclear Waste
Management, disposal and governance
Klaus-Jürgen Röhlig

Chapter 9

Decision-making in the presence of uncertainty: the safety case as a tool supporting the development of a deep geologic repository

Klaus-Jürgen Röhlig

9.1 Safety-related statements, assessments, and reviews that support decisions in a stepwise repository programme

Safety, in particular, long-term safety (i.e. safety after repository closure), is the obvious and ultimate goal of constructing a deep (or any other) disposal facility. As described in chapter 8 'Geologic ('deep') disposal of high-level and other long-lived waste: host rocks, concepts, current international status,' there are various strategies for achieving this goal, all of which focus on the aims described hereinafter:

> 'Disposal options are designed to contain the waste by means of passive engineered and natural features and to isolate it from the accessible biosphere to the extent necessitated by the associated hazard.
>
> [...]
> The specific aims of disposal are:
> (a) To contain the waste;
> (b) To isolate the waste from the accessible biosphere and to reduce substantially the likelihood of, and all possible consequences of inadvertent human intrusion into the waste;
> (c) To inhibit, reduce and delay the migration of radionuclides at any time from the waste to the accessible biosphere;
> (d) To ensure that the amounts of radionuclides reaching the accessible biosphere due to any migration from the disposal facility are such that possible radiological consequences are acceptably low at all times.' (IAEA 2011a)

doi:10.1088/978-0-7503-3095-4ch9

'Containment denotes all methods or physical structures designed to prevent or control the release and the dispersion of radioactive substances. Isolation of the waste from the accessible biosphere substantially reduces the likelihood of inadvertent human intrusion into the waste and its consequences.' (IAEA 2012)

The challenge is not only to construct a facility that meets these aims, but also to demonstrate their fulfilment in a credible way. It is likely (and has happened frequently in the past) that such a demonstration will be heavily contested by stakeholders, for various reasons: safety is a concern for all interested actors, and the idea of demonstrating it over time frames that exceed the usual horizons of human perceptions by far might be perceived as hubris. Moreover, the opponents of a facility or a programme might see issues related especially to long-term safety as the most vulnerable points of such a programme, and their counter-arguments as the most promising ones. This, as well as the perhaps overly naive approach taken by some proponents, might, at least in part, explain the early failures of many disposal programmes (see part II, chapter 12 'The politization and politics of nuclear waste: a socio-technical history' and chapter 13 'Role of stakeholder involvement in the implementation of radioactive waste management projects'). The nature of these failures is perhaps described most instructively in the report of the so-called 'Seaborn Panel' (CEAA 1998), which evaluated a disposal project for spent nuclear fuel presented by Atomic Energy of Canada Ltd (AECL) as follows:

– 'From a technical perspective, the safety of the AECL concept has, on balance, been adequately demonstrated for a conceptual stage of development, but from a social perspective, it has not.
– As it stands, the AECL concept for deep geological disposal has not been demonstrated to have broad public support. The concept in its current form does not have the required level of acceptability to be adopted as Canada's approach for managing nuclear fuel wastes.'

The safety assessment, i.e. the 'systematic analysis of the hazards associated with a geological disposal facility, and the ability of the site and design to provide the safety functions and meet technical requirements' (NEA 2012), was initially focussed on calculating estimates of the long-term radiological impact caused by the migration of radionuclides into the accessible biosphere for various potential evolutions of the system called 'scenarios' (NEA 1991a). Numerical (computer) simulations of groundwater flow and the resulting contaminant migration played a central role in such assessments, and considerable efforts were made to verify and validate the computer models used for such simulations (NEA 1991b, 1995).

Since then, it has increasingly been recognised that a more holistic approach to safety is needed, from both technical and societal perspectives. It has been acknowledged that disposal programmes will need considerably more time than initially anticipated, and that they require much better provision for the societal context they are embedded in. Increasingly, a 'Stepwise Approach to Decision Making for

Long-term Radioactive Waste Management' (NEA 2004a), with the possibility of reversing actions if necessary, has been considered advisable. The typical stages of a disposal programme include conceptualisation, site selection and investigation, construction, operation (starting when the facility receives the first waste and thus becomes a nuclear facility), closure, and the post-closure period. Site selection in itself might require several substages, during which the range of regions, areas, or sites considered is narrowed down (IAEA 2011b). Obviously, such stages reach points at which decisions of different types have to be taken concerning the repository concept, the site(s) to be considered or investigated, and the start of construction, operation, or closure of a facility. The types of decisions not only include the passing of laws, other regulations, or licensing, but also internal decisions within organisations, e.g. those concerning a certain facility design.

The type of decision that has to be taken at a particular stage and the actors to be involved depend very much on the national political and legal framework. However, they all have the common factor that safety is an issue of utmost importance—the decisions have to be safety-driven or at least safety-informed. The tool that is intended to provide the decision-makers and other concerned or interested actors with the necessary safety-relevant information is the Safety Case, a 'collection of scientific, technical, administrative and managerial arguments and evidence in support of the safety of a disposal facility, covering the suitability of the site and the design, construction, and operation of the facility, the assessment of radiation risks and assurance of the adequacy and quality of all of the safety related work associated with the disposal facility' (IAEA 2012). Safety Assessment stands at the core and forms an integral part of the Safety Case.

According to IAEA (2012), the 'concept of developing a safety case for disposal facilities [...] is used in many States. The terminology used is different, though, in some States.' Named examples are:

- 'total system performance analysis' (United States of America)
- 'dossier de sûreté' (France)
- 'Sicherheitsnachweis' (Switzerland)
- 'estudio de seguridad' (Spain)
- 'Sicherheitsbericht' (safety report, not yet named in the IAEA Guide) is being used in Germany according to recently adopted regulation (EndlSiAnfV 2020; the term 'Sicherheitsnachweis' was used previously).

Since the Safety Case has to inform relevant programme decisions, it has to provide information about what has been achieved at a certain stage of a repository programme, but also about what has not (yet) been achieved and what the important uncertainties and potential obstacles for the programme are. It should also contain proposals about how to address these uncertainties e.g. by means of site investigation, research, and development. The decision to be taken at a certain stage of a programme will then include these means, e.g. by defining a research and development strategy. The results of these activities will be fed into the Safety Case to be presented at the next stage. In other words, the Safety Case will evolve over time together with the disposal programme by taking acquired information into account and, of course, by possibly formulating new questions to be addressed at the next stage.

If a Safety Case is meant to support a legal or licensing decision, it is important that the decision-makers (policy makers, licensing authorities, etc.) are independent from the producer of the Safety Case (the 'implementer'), as required by the EU directive (EU 2011). Moreover, the decision-makers should have access to independent technical expertise so that they can review the Safety Case. Involving international expertise, as, for example, described in the guidelines for peer reviews carried out by the OECD's Nuclear Energy Agency (NEA 2005), can be particularly helpful, no matter whether the Safety Case is to support legal or licensing decisions or, by contrast, 'internal' decisions to be made by the implementer.

As stakeholder involvement and participation are crucial in such decision-making (see part II chapter 13 'Role of stakeholder involvement in the implementation of radioactive waste management projects'), and because safety is a central issue in the discourse, it is furthermore important that the content of the safety case is transparently communicated in a manner that is accessible to different audiences. Details and examples were addressed, for example, by the NEA (2017), but the issue deserves further investigation and development, some of which the NEA undertook in 2021 (NEA 2021a).

9.2 Safety case: evolution and elements—status and perspectives

An example of the evolution of a Safety Case as described above is the French high-level waste (HLW) disposal programme, which comprises such a sequence of evolving Safety Cases. In 1991, the French parliament passed a law ('Loi Bataille')[1], in which a time frame of 15 years was defined for the investigation of three 'research avenues' for HLW management, namely partitioning and transmutation, reversible or irreversible disposal in deep geologic formations, and long-term storage. Thus, the 'Loi Bataille' became, amongst other things, the starting point for the evolution of the French HLW disposal Safety Case:

- 1996: In order to investigate disposal in argillaceous rock, the construction of an underground research laboratory (URL) at the Bure site was applied for in 1996 and licensed in 1998. According to Lebon (2009), a preliminary safety assessment formed an important basis for this **licensing decision** and can therefore be seen as a first in a series of evolving safety cases.
- 2001: In order to prepare for the legal decision(s) to be made according to the 15-year time frame set by the 'Loi Bataille' in 2006 and in order to present design options and to **test the methodology** to be applied, the French implementing organisation prepared a preliminary Safety Case called 'Dossier 2001 Argile' (Andra 2001a, 2001b).
- 2005: As explained above, for 2006, according to the 'Loi Bataille,' a decision on the way forward was needed. With the 'Dossier 2005 Argile' (Andra 2005a) a Safety Case was provided to support the related **legislative process**; based on the

[1] Loi n° 91–1381 du 30 décembre 1991 relative aux recherches sur la gestion des déchets radioactifs.

geoscientific information acquired at the Bure URL, a concept for retrievable disposal was presented and its safety was assessed in order to support the claim that a disposal concept in the vicinity of the URL was feasible. Also, a 250 km^2 area around the URL was identified, in which the repository was to be sited. In addition, a less detailed 'Dossier 2005 Granite' (Andra 2005b) was presented which was not based on site-specific, but on generic data. The legal procedure lead to a new law requiring, *inter alia*, further studies in order to identify a site and submit a license application for reversible HLW disposal in 2015, and the development of a 'National Radioactive Materials and Waste Management Plan' (for a summary of the law, see Andra 2016e).

- 2009: In order to **support siting**, Andra submitted a report in which it '... proposed to the French Government a 30 km^2 underground zone (ZIRA, or zone of interest for detailed reconnaissance) within the 250 km^2 zone defined in 2005' (Andra 2016b).
- 2016: For various reasons, including the need to account for input from public participation concerning the implementation of an additional 'industrial pilot phase,' Andra did not reach the target of preparing a license application in 2015. Instead, as a **preliminary step in advance of the actual application** a 'Safety Options Report' (Andra 2014, 2016a, 2016b, 2016c, 2016d) was prepared in order 'to prepare for the examination of the construction licence application' (Andra 2016b) according to proposals made by Andra and expectations expressed by the licensing authority, Autorité de sûreté nucléaire (ASN).

Obviously, these Safety Cases, despite the fact that each of them evolved from its predecessor(s), vary considerably in purpose, scope, and intended audience. In addition, the role of the licensing authority ASN varied. ASN was, however, involved in all the steps described above and formulated opinions and advice, although it did not carry out a formal review of each of the steps. In doing so, it was supported by its Technical Support Organisation, Institut de radioprotection et de sûreté nucléaire (IRSN). In addition, the advisory body Groupe permanent d'experts pour les déchets (GPD) provided its own statements and opinions. Public participation played a role in the development of the project and in at least one case, prompted an important modification: the planning of the 'industrial pilot phase' mentioned above. Moreover, the 2001, 2005, and 2016 dossiers underwent peer reviews by international expert teams, the former two on behalf of Andra's supervisory ministries and coordinated by OECD's NEA (NEA 2003, 2006), the latter on behalf of the ASN and coordinated by the IAEA (2017). The recommendations of these reviews were taken into account in the succeeding stages of the programme.

From the above, it is obvious that a Safety Case evolves and becomes increasingly mature: the results of ongoing research, development, and site investigation are taken into account, but of course also cause new questions to be addressed. At each decision point, uncertainties have to be acknowledged and the questions have to be asked: Is it prudent to move forward in the presence of these uncertainties? Is there a need for programme changes? Which of the uncertainties are to be resolved later, and, if so, when and how? Which uncertainties are acceptable, or perhaps unavoidable? These questions have to be asked at each decision point, but of course they are most crucial when it comes to legal or licensing decisions.

Within the Safety Case, uncertainties have to be identified to the extent possible. The first questions to be asked are those about their magnitude, their relevance for safety, and about the quality or reliability of statements which can be made about magnitude and relevance. Depending on the outcome, one then has to consider options for addressing ('managing') the uncertainty (Eckhardt 2021). Such options include avoiding the uncertainty e.g. by siting or design choices for which the uncertainty is insignificant, mitigating its impact on safety e.g. by 'over-engineering', or reducing it by further research or site investigation efforts (Vigfusson *et al* 2007).

While this continuous treatment of uncertainties leading to increasing maturity is one factor in the evolution of the Safety Case, there is more: Disposal programmes evolve and mature as well, along with changing necessities and thus changing emphasis. Moreover, the state of the art that is recognised internationally evolves in consonance with national developments.

In the 1990s, considerations about the (post-closure) Safety Case aimed to overcome the confidence problems linked to Safety Assessments solely focussed on the estimation of radiation exposures caused by radionuclide migration for various scenarios using computer simulations and the attempts to verify and validate the models used (NEA 1991a, 1991b, 1995, see previous subchapter). These considerations focussed on the confidence in the arguments used when claiming long-term safety and resulted in the request for a more holistic approach that would integrate site investigation, scientific, design and engineering considerations, and safety assessment (sometimes termed 'performance assessment'; the dividing line between the two is not always clear) as well as other (possibly qualitative) arguments. Methodological developments for such a 'Post-closure Safety Case' that relied on and inspired approaches taken nationally and research undertaken at the EU level spanned more than a decade (NEA 1997, 1999, 2000, 2002, 2004b, 2008, 2009, EC 2011). According to NEA (2013), a Post-closure Safety Case should contain at least:

- A clear statement of the safety case context and purpose, in particular, a description of the decision to be supported.
- A safety strategy, i.e. a description of 'the high-level approach adopted to achieve safe disposal' (the safety concept), including 'an overall management strategy, a siting and design strategy and an assessment strategy'.
- A description of the assessment basis, i.e. 'the collection of information and analysis tools supporting the safety assessment' including 'an overall description of the disposal system [...], the scientific and technical data and understanding relevant to the assessment of system safety, and the assessment methods, models, computer codes and databases'.
- A safety assessment based on 'a methodological approach to the numerous processes, features and other technical issues that need to be systematically evaluated [...] using either mathematical analyses or more qualitative arguments, under a range of evolution scenarios for the disposal system. Robustness of the safety case is strengthened by the use of multiple lines of

evidence leading to complementary safety arguments that can compensate for shortcomings in any single argument. One approach to creating a basis for a conceptual understanding of a geological system and its important operative processes is to collect data into a 'geosynthesis'—a data model that illustrates data as distributed over the entire geological system [...]. More generally, the conceptual understanding of a repository system may also be established through a series of process reports which present the thermal, hydraulic, mechanical and chemical process information of the engineered and geological components of the repository system over the time frame of concern. This gives both an indication of how well the system has been characterised, and illustrates the integration of the science, the design and the modelling that simulates system evolution with time.'

- A statement of confidence needed to 'justify a positive decision to proceed to the next stage of planning or implementation [...] based on the analyses and arguments developed and the supporting evidence gathered. The statement of confidence recognises the existence of open issues and uncertainties, and perspectives about how they can be addressed in the next step(s). The audience of the safety case must decide whether it believes the reasoning that is presented is adequate, and on that basis whether, or to which extent, it shares the confidence of the safety case author.'

The requirements for the production and documentation of the Safety Case are laid down in the IAEA Safety Standards (in particular IAEA 2012), which recognise that its scope will extend beyond the scientific and technical issues related to post-closure safety and include issues such as safe construction, the operation and closure of a facility, and management issues. The degree to which safety during the construction, operation, and closure of a repository, technical feasibility, and their interplay with long-term safety are to be addressed in a Safety Case will depend on the maturity of the project; for example, in the example from France described above, the focus given to these issues increases from step to step. In the 2016 Safety Case, a volume is dedicated to operational safety (Andra 2016a), while another contains the abovementioned elements that are important for post-closure issues (Andra 2016b). In accordance with the purpose of the Safety Case, further volumes are dedicated to retrievability (Andra 2016c) and to a 'master plan for the operation' (Andra 2016d) as well as to a 'Development plan for Cigéo project components' (Andra 2014).

Similar observations can be made for other national projects: technical feasibility and operational safety become increasingly important considerations in Safety Cases (NEA 2008, 2015a, 2015b). Other issues that emerge from the Safety Case methodology are the relationship to communication, stakeholder involvement, and public participation (NEA 2008, 2014, 2017, 2021a, 2021b), information and knowledge management (NEA 2018), and issues related to human and organisational factors in general (Eckhardt 2021). Developing and establishing approaches towards an 'Electronic' or 'Digital Safety Case' is currently being considered (cf. eg. https://www.oecd-nea.org/jcms/pl_53147/expert-group-on-a-data-and-information-management-strategy-for-the-safety-case-egssc) and might help addressing such emerging issues.

Bibliography

Andra (Agence nationale pour la gestion des déchets radioactifs) 2001a Dossier 2001 *Argile, Synthesis Report, Part* A Châtenay-Malabry, France

Andra (Agence nationale pour la gestion des déchets radioactifs) 2001b Dossier 2001 *Argile, Synthesis Report, Part* B Châtenay-Malabry, France

Andra (Agence nationale pour la gestion des déchets radioactifs) 2005a Dossier 2005 *Argile. Evaluation of the Feasibility of a Geological Repository in an Argillaceous Formation. Meuse/ Haute-Marne site* https://international.andra.fr/sites/international/files/2019-03/3-%20Dossier%202005%20Argile%20Synthesis%20-%20Evaluation%20of%20the%20feasibility%20of%20a%20geological%20repository%20in%20an%20argillaceous%20formation_0.pdf

Andra (Agence nationale pour la gestion des déchets radioactifs) 2005b Dossier 2005 *Granite. Assets of Granite Formations for Deep Geological Disposal* https://international.andra.fr/sites/international/files/2019-03/7-%20Dossier%202005%20Granite%20Synthesis%20-%20Assets%20of%20granite%20formations%20for%20deep%20geological%20disposal_0.pdf

Andra (Agence nationale pour la gestion des déchets radioactifs) 2014 *Plan de développement des composants du projet Cigéo—Déclinaison suivant l'échelle TRL (ISO 16290:2013)* CG.PDD. ADPG.14.0031/A

Andra (Agence nationale pour la gestion des déchets radioactifs) 2016a *Safety Options Report—Operating Part (DOS-Expl)* CG-TE-D-NTE-AMOA-SR1-0000-15–0060 https://international.andra.fr/sites/international/files/2019-03/Safety%20Options%20Report%20-%20Operations_2.pdf

Andra (Agence nationale pour la gestion des déchets radioactifs) 2016b *Safety Options Report—Post-closure Part (DOS-AF)* CG-TE-D-NTE-AMOA-SR2-0000-15–0062 https://international.andra.fr/sites/international/files/2019-03/Safety%20Options%20Report%20-%20Post-Closure_1.pdf

Andra (Agence nationale pour la gestion des déchets radioactifs) 2016c *Dossier d'Options Techniques de Récupérabilité (DORec)* CG-TE-D-NTE-AMOA-RV0-0000-15–0059/A https://andra.fr/sites/default/files/2018-11/CG-TE-D-NTE-AMOA-RV0-0000-15–0059-A%20DOREC.pdf

Andra (Agence nationale pour la gestion des déchets radioactifs) 2016d *Proposition de Plan Directeur pour l'Exploitation de Cigéo* CG-TE-D-NTE-AMOA-SDR-0000–15-0063/A https://andra.fr/sites/default/files/2018-08/proposition%20PDE%20avril%202016.pdf

Andra (Agence nationale pour la gestion des déchets radioactifs) 2016e *Radioactive Materials and Waste Planning Act of 28 June 2006* Consolidated version established by Andra https://international.andra.fr/sites/international/files/2019-03/loi%202016.pdf

CEAA (Canadian Environmental Assessment Agency) 1998 *Nuclear Fuel Waste Management and Disposal Concept: Report of the Nuclear Fuel Waste Management and Disposal Concept Environmental Assessment Panel* (Ottowa: Public Works and Government Services Canada)

Eckhardt A 2021 *Sicherheit angesichts von Ungewissheit. Ungewissheiten im Safety Case (in German, Safety in the Presence of Uncertainty. Uncertainties in the Safety Case)* Transens Report No. 1 https://doi.org/10.21268/20210412-0

EndlSiAnfV (Endlagersicherheitsanforderungsverordnung) 2020 *Verordnung über Sicherheitsanforderungen an die Endlagerung hochradioaktiver Abfälle (in German, Ordinance on Safety Requirements for the Disposal of High-level Radioactive Waste)* http://gesetze-im-internet.de/endlsianfv/

EC (European Commission) 2011 *Performance Assessment methodologies in application to guide the development of the safety case* Project summary report http://ip-pamina.eu/downloads/pamina.summaryweb.pdf

EU 2011 Council Directive 2011/70/EURATOM 2011 of 19 July establishing a Community framework for the responsible and safe management of spent fuel and radioactive waste *Off. J. Eur. Union* L 199/48 https://eur-lex.europa.eu/legal-content/EN/TXT/?uri=celex%3A32011L0070

IAEA (International Atomic Energy Agency) 2011a *Disposal of Radioactive Waste. Specific Safety Requirements* Specific Safety Requirements, IAEA Safety Standards Series No. SSR-5 https://www-pub.iaea.org/MTCD/Publications/PDF/Pub1449_web.pdf

IAEA (International Atomic Energy Agency) 2011b *Geological Disposal Facilities for Radioactive Waste* Specific Safety Guide, IAEA Safety Standards Series No. SSG-14 https://www-pub.iaea.org/MTCD/Publications/PDF/Pub1483_web.pdf

IAEA (International Atomic Energy Agency) 2012 *The Safety Case and Safety Assessment for the Disposal of Radioactive Waste* Specific Safety Guide, IAEA Safety Standards Series No. SSG-23 https://www-pub.iaea.org/MTCD/Publications/PDF/Pub1553_web.pdf

IAEA (International Atomic Energy Agency) 2017 *An International Peer Review of the Safety Options Dossier of the Project for Disposal of Radioactive Waste in Deep Geological Formations (Cigéo)* Final report of the IAEA international review team (Vienna: IAEA) https://www-pub.iaea.org/MTCD/Publications/PDF/CIG_FRA_web.pdf

Lebon P 2009 Overview of the disposal feasibility assessment in meuse/haute-marne: from the preliminary geoscientific survey to the safety case in approaches and challenges for the use of geological information in the safety case for deep disposal of radioactive waste *Proc. of the AMIGO-3 Workshop (Nancy, France, 15–17 April 2008)* https://oecd-nea.org/rwm/reports/2009/AMIGO-3/

NEA (Nuclear Energy Agency) 1991a *Disposal of Radioactive Waste: Review of Safety Assessment Methods* (Paris: OECD)

NEA (Nuclear Energy Agency) 1991b *GEOVAL-1990: Safety Assessment of Radioactive Waste Repositories- Symposium on Validation of Geosphere Flow and Transport Models (Stockholm, 14-17 May 1990) (English and French Edition)* (Paris: OECD)

NEA (Nuclear Energy Agency) 1995 *GEOVAL '94 : Validation Through Model Testing: Proc. of an NEA/SKI Symp. (Paris, France, 11–14 October 1994)* (Paris: OECD)

NEA (Nuclear Energy Agency) 1997 *Learnt from Ten Performance Assessment Studies* OECD https://oecd-nea.org/jcms/pl_13140/lessons-learnt-from-ten-performance-assessment-studies

NEA (Nuclear Energy Agency) 1999 *Confidence in the Long-term Safety of Deep Geological Repositories. Its Development and Communication* OECD https://oecd-nea.org/jcms/pl_13274/confidence-in-the-long-term-safety-of-deep-geological-repositories?details=true

NEA (Nuclear Energy Agency) 2000 *Regulatory Reviews of Assessments of Deep Geologic Repositories* OECD https://oecd-nea.org/jcms/pl_13358/regulatory-reviews-of-assessments-of-deep-geologic-repositories/evaluation-des-depots-geologiques-profonds-dans-un-contexte-reglementaire?details=true

NEA (Nuclear Energy Agency) 2002 *Establishing and Communicating Confidence in the Safety of Deep Geologic Disposal* OECD https://oecd-nea.org/jcms/pl_13638/establishing-and-communicating-confidence-in-the-safety-of-deep-geologic-disposal/etablir-et-faire-partager-la-confiance-dans-la-surete-des-depots-en-grande-profondeur?details=true

NEA (Nuclear Energy Agency) 2003 *The French R&D Programme on Deep Geological Disposal of Radioactive Waste. An International Peer Review of the 'Dossier 2001 Argile'* OECD https://oecd-nea.org/jcms/pl_13796/the-french-r-d-programme-on-deep-geological-disposal-of-radioactive-waste

NEA (Nuclear Energy Agency) 2004a *Stepwise Approach to Decision Making for Long-term Radioactive Waste Management. Experience, Issues and Guiding Principles* OECD, NEA No. 4429 https://oecd-nea.org/jcms/pl_13790/stepwise-approach-to-decision-making-for-long-term-radioactive-waste-management?details=true

NEA (Nuclear Energy Agency) 2004b *Post-closure Safety Case for Geological Repositories. Nature and Purpose* OECD, NEA No. 3679 https://oecd-nea.org/jcms/pl_13678/post-closure-safety-case-for-geological-repositories

NEA (Nuclear Energy Agency) 2005 *International Peer Reviews in the Field of Radioactive Waste: Questionnaire on Principles and Good Practice for Safety Cases* OECD, NEA/RWM/PEER (2005)2 https://oecd-nea.org/jcms/pl_18276/international-peer-reviews-in-the-field-of-radioactive-waste-questionnaire-on-principles-and-good-practice-for-safety-cases/revues-internationales-par-des-pairs-dans-le-domaine-des-dechets-radioactifs-questionnaire-sur-les-principes-et-bonnes-pratiques-concernant-les-dossiers-de-surete

NEA (Nuclear Energy Agency) 2006 *An International Peer Review of the 'Dossier 2005 Argile' Concerning Disposal in the Callovo-Oxfordian Formation* OECD, NEA/RWM/PEER(2006)1 https://oecd-nea.org/jcms/pl_18334/an-international-peer-review-of-the-dossier-2005-argile-concerning-disposal-in-the-callovo-oxfordian-formation

NEA (Nuclear Energy Agency) 2008 *Safety Cases for Deep Geological Disposal of Radioactive Waste: Where Do We Stand? (Symposium Proceedings, Paris, France 23–25 January 2007)* OECD, NEA No. 6319 https://oecd-nea.org/upload/docs/application/pdf/2020-11/ne6319-safety.pdf

NEA (Nuclear Energy Agency) 2009 *International Experiences in Safety Cases for Geological Repositories (INTESC)* OECD, NEA No. 6251 https://oecd-nea.org/jcms/pl_14264/international-experiences-in-safety-cases-for-geological-repositories-intesc-cd-rom?details=true

NEA (Nuclear Energy Agency) 2012 *Methods for Safety Assessment of Geological Disposal Facilities for Radioactive Waste. Outcomes of the MeSA Initiative* OECD, NEA No. 6923 https://oecd-nea.org/jcms/pl_14608/methods-for-safety-assessment-of-geological-disposal-facilities-for-radioactive-waste

NEA (Nuclear Energy Agency) 2013 *The Nature and Purpose of the Post-closure Safety Cases for Geological Repositories* OECD, NEA/RWM/R(2013)1 https://oecd-nea.org/rwm/reports/2013/78121-rwn-sc-brochure.pdf

NEA (Nuclear Energy Agency) 2014 *The Safety Case for Deep Geological Disposal of Radioactive Waste: 2013 State of the Art. Symposium Proceedings, 7–9 October 2013, Paris, France* OECD, NEA/RWM/R(2013)9 https://oecd-nea.org/rwm/reports/2013/78121-rwn-sc-brochure.pdf

NEA (Nuclear Energy Agency) 2015a *Interim Report on Fire Risk Management* OECD, NEA/RWM/R(2015)6 https://oecd-nea.org/jcms/pl_55414/interim-report-on-fire-risk-management

NEA (Nuclear Energy Agency) 2015b *Analysis of a Questionnaire on Hazards Associated with Coactivities in Deep Geological Repositories 2013–15* OECD, NEA/RWM/R(2015)7 https://oecd-nea.org/jcms/pl_55404/analysis-of-a-questionnaire-on-hazards-associated-with-co-activities-in-deep-geological-repositories-2013–2015

NEA (Nuclear Energy Agency) 2017 *Communication on the Safety Case for a Deep Geological Repository* OECD, NEA No. 7336 https://oecd-nea.org/jcms/pl_15032/communication-on-the-safety-case-for-a-deep-geological-repository

NEA (Nuclear Energy Agency) 2018 *Managing Information and Requirements in Geological Disposal Programmes* OECD, NEA/RWM/R(2018)2 https://oecd-nea.org/jcms/pl_31187/managing-information-and-requirements-in-geological-disposal-programmes

NEA (Nuclear Energy Agency) 2021a *Managing Uncertainty in Siting and Implementation—Creating a Dialogue Between Science and Society* OECD https://oecd-nea.org/jcms/pl_59339/2nd-igsc-fsc-workshop-managing-uncertainty-in-siting-and-implementation-9-october-2019

NEA (Nuclear Energy Agency) 2021b *Current Understanding and Future Direction for the Geological Disposal of Radioactive Waste. Symposium Proceedings, 10–11 October 2018, Rotterdam, Netherlands* OECD, in preparation

Vigfusson J, Maudoux J, Raimbault P, Röhlig K-J and Smith R E 2007 *European Pilot Study on the Regulatory Review of the Safety Case for Geological Disposal of Radioactive Waste. Case study: Uncertainties and their Management* https://grs.de/sites/default/files/pdf/pilotstudie_radioactive_waste_0.pdf

Part II

Nuclear waste management and society

IOP Publishing

Nuclear Waste
Management, disposal and governance
Klaus-Jürgen Röhlig

Chapter 10

Legal and regulatory issues

Hans Wanner

10.1 The International Atomic Energy Agency, IAEA

The International Atomic Energy Agency (IAEA) is an international organisation established in 1957. The main objective of the IAEA is to promote the peaceful use of nuclear energy, and to inhibit the use of nuclear energy for military purposes (IAEA 1989). The IAEA is headquartered in Vienna (Austria) and reports annually to the General Assembly of the United Nations. Most UN members and the Holy See are Member States of the IAEA (IAEA 2021a).

The IAEA has established a comprehensive series of safety standards that are used as a reference for the legal and regulatory systems in its member States. The IAEA also offers a large range of peer review services that allow member states to check the correct implementation of these standards in their countries. Strictly speaking, the IAEA safety standards are not legally binding for the member states. However, the nuclear legislation of many countries requires periodical invitations of IAEA International Regulatory Review Service (IRRS) missions. Such missions check the legal and regulatory system against IAEA safety standards and issue recommendations for improvement in cases of lacking or insufficient implementation. So-called follow-up missions within four years after the initial IRRS mission have the purpose of checking the implementation of the recommendations. From this viewpoint, we can say that the IAEA safety standards are binding in an indirect way. Further information is provided in the paragraph below on IRRS missions.

In the area of radioactive waste management, there is another effective instrument that checks the implementation of the IAEA safety standards: the Joint Convention on the Safety of Spent Fuel Management and on the Safety of Radioactive Waste Management, in short, the 'Joint Convention.' The countries that have signed and ratified the Joint Convention are obliged to report on the state of implementation of the articles of the Joint Convention in National Reports every three years. Further information is provided in the paragraph below on the Joint Convention.

The IAEA Safety Standards

The Safety Standards of the IAEA are grouped into a number of hierarchical levels (IAEA 2021b):

1. Safety Fundamentals
2. General Safety Requirements
3. Specific Safety Requirements
4. General Safety Guides
5. Specific Safety Guides

The Safety Fundamentals SF-1 (IAEA 2006) provide the basis for the IAEA Safety Standards. Related requirements are established in the Safety Requirements publications. Guidance on meeting these requirements is provided in the related Safety Guides.

Ten principles are at the core of the Safety Fundamentals, which are worth mentioning:

Principle 1: Responsibility for safety
Principle 2: Role of government
Principle 3: Leadership and management for safety
Principle 4: Justification of facilities and activities
Principle 5: Optimization of protection
Principle 6: Limitation of risks to individuals
Principle 7: Protection of present and future generations
Principle 8: Prevention of accidents
Principle 9: Emergency preparedness and response
Principle 10: Protective actions to reduce existing or unregulated radiation risks

The Recommendations are subordinate to the Fundamentals. Likewise, the Guides are subordinate to the Recommendations. All of them have the overarching term 'Safety Standards' in common.

International Peer Review Service (IRRS) missions

The guideline for IRRS missions (IAEA 2018) gives the following description of the purpose and conduct of such missions in its chapter 1 (Introduction): 'The Integrated Regulatory Review Service was developed to support the application of the IAEA Safety Fundamentals, safety requirements and guides and to enhance the effectiveness of national regulatory infrastructures for safety. It requires a Member State to conduct a self-assessment of its regulatory infrastructure against the IAEA safety standards, in accordance with the IAEA Self-Assessment Methodology. This is then followed by an IAEA-coordinated peer review conducted by a team of international reviewers, led by a senior regulator from a Member State. The review is structured to lead to the identification of areas for improvement and the formulation, by the host country, of an action plan to address identified deficiencies. IRRS missions provide an opportunity for the host country and peer review team members to share experience regarding both regulatory technical and policy matters. Discussions conducted during IRRS missions take into account issues identified during both the States' self-assessments and the IRRS Team's review of regulatory technical areas.'

IRRS missions are usually requested by a formal letter from the host country to the IAEA about two years in advance of the proposed mission date to allow sufficient time for the self-assessment and the preparation of the mission. The IAEA designates one of its staff members as the IAEA Coordinator and selects the IRRS Team Leader and Deputy Team Leader. These are senior regulators from IAEA Member States. The other members of the IRRS review team, the IRRS Reviewers, are experienced regulators recruited from IAEA Member States in consultation with the host country.

The most important work during the preparation phase is the performance of a self-assessment, by which the host country demonstrates and proves its compliance with the IAEA safety standards. This is a time-consuming process, which takes at least a year to be accomplished. In general, the host country will identify deficiencies during this process, and will prepare an action plan to address the recognised shortcomings in the national legal, governmental, and regulatory framework for safety. The result is a comprehensive report that forms the core of the so-called Advance Reference Material, which is submitted to the IAEA. This is the most important material used by the IRRS reviewers to prepare for the IRRS mission.

The IRRS mission is conducted in the host country and usually lasts for two weeks. It starts with an entrance meeting in the presence of the senior management of the host country's regulatory body and relevant government officials, and ends with an exit meeting at which the IRRS review team presents its findings. During the mission, the IRRS reviewers conduct interviews with all bodies involved in the regulation of nuclear, radiation, waste, and transport safety, including technical support organisations, research institutes, and regulated organisations, such as nuclear power plants and waste management companies. Certain interviews and direct observations of regulatory activities, such as inspections, are carried out at the sites of the regulated organisations. The IRRS reviewers make detailed notes of all relevant observations, which form the basis for the IRRS report. All the issues identified are brought to the attention of the other IRRS team members. The IRRS team discusses the issues and agrees on the formulation of recommendations, suggestions, and good practices.

Recommendations are made when the arrangements required to meet the IAEA Safety Requirements are either missing, incomplete, or inadequately implemented. Suggestions are made when opportunities for improvement are identified and should be considered. They often have a basis in the IAEA Safety Guides. A good practice is identified in recognition of an outstanding element which is superior to those generally observed elsewhere and worthy of the attention of other regulatory bodies.

The recommendations, suggestions, and good practices are the essential results of the IRRS mission and are documented in the IRRS mission report. Until the nuclear accident at Fukushima Daiichi, IRRS mission reports used to be restricted unless the host country decided to make them public. Today, however, IRRS mission reports are released by default, unless the host country specifically requests that they remain restricted. This was an important change in favour of transparency.

Within four years after an IRRS mission, a follow-up mission is undertaken to assess the implementation of the recommendations and suggestions in the period since the initial review mission.

Many countries have legal requirements to carry out periodic international peer reviews, such as those performed by the IRRS, often with interval requirements such as ten years, as, for example, required by the European Council Directive 2014/84/EURATOM of 8 July 2014 (EU 2014).

Other Peer Review Missions of the IAEA

The IAEA offers quite a number of additional peer review missions (IAEA 2021c). Some of them are relevant to radioactive waste management, in particular, this is the case for ARTEMIS (the Integrated review service for radioactive waste and spent fuel management, decommissioning, and remediation). The process of such missions is analogous to the IRRS mission. Like the IRRS missions, all these peer reviews provide valuable outside views and offer excellent opportunities for hosts and peers to exchange experience and to learn from each other.

The Joint Convention on the Safety of Spent Fuel Management and on the Safety of Radioactive Waste Management (Joint Convention)

A different tool for checking the implementation of the IAEA Safety Standards, in particular in the area of spent fuel and radioactive waste management, is the Joint Convention (IAEA 2021d). The difference lies in the fact that countries (so-called contracting parties) that have signed and ratified the Joint Convention are obliged to report, at intervals of three years, on the status of implementation of the Articles of the Convention. The essence of these Articles can be outlined by taking the example of the management of radioactive waste. The countries have to report on the following issues:

1. General safety requirements
2. Existing facilities and past practices
3. Siting of proposed facilities
4. Design and construction of facilities
5. Assessment of the safety of facilities
6. Operation of facilities
7. Institutional measures after closure (of the repository)

This is done through the submission of national reports by a given deadline. The national reports are made accessible to all other contracting parties, which are invited to study them and to address questions and comments. The core of the review process is the review meeting in Vienna, during which each country will present its national report, address the questions raised, and will receive assessments in the form of good practices, areas of good performance, suggestions, and challenges formulated by the other participating countries.

This review tool is quite effective, because the countries (as they are contracting parties of the Joint Convention) have a high-level interest in compliance and improvement from one review meeting to the next.

10.2 The Western European Nuclear Regulators' Association, WENRA

The Western European Nuclear Regulators' Association (WENRA), established in 1999, is an association of the Heads of the nuclear regulatory bodies in Europe (WENRA 2021b). Its currently 18 members are regulators from European countries with a nuclear power programme, and its 14 observers (including so-called associated members) are from European Countries without a nuclear programme, or from countries with a nuclear programme that are outside the European Union. The latter may apply for WENRA member status if they can demonstrate compliance with the WENRA safety standards. The main purposes of WENRA are to harmonize the safety regulations in member countries, exchange experience, and discuss significant safety issues.

WENRA issues safety standards, the so-called Safety Reference Levels (sometimes just called Reference Levels). They are proposed by the WENRA technical working groups and obtained by analysing individual national approaches, comparing them with the IAEA Safety Standards, identifying any differences, and proposing a way forward to eliminate the differences. The proposals are based on best practices among the most advanced requirements for nuclear facilities, and are submitted to the WENRA plenary for approval. The current WENRA working groups are the Reactor Harmonisation Working Group (RHWG) and the Working Group on Waste and Decommissioning (WGWD). WENRA also issues guidance documents and reports on specific issues, as well as specific recommendations. WENRA members commit to implementing the WENRA safety standards and recommendations in their countries.

In the working area of the WGWD, WENRA issued Safety Reference Levels on spent fuel storage in 2014 (WENRA 2014a), on waste disposal facilities in 2014 (WENRA 2014c), on decommissioning in 2015 (WENRA 2015), on waste treatment and conditioning in 2018 (WENRA 2018), as well as an umbrella document on interfaces and interdependencies in 2017 (WENRA 2017).

In the working area of the RHWG, WENRA issued several versions of Safety Reference Levels, the first in 2006 (WENRA 2006), then in 2008 (WENRA 2008), in 2014 (an update due to lessons learned from the Fukushima nuclear accident, WENRA 2014b), and most recently in 2021 (WENRA 2021a). Additional WENRA reports and guidance documents have been worked out and prepared by the RHWG.

The implementation of the WENRA standards and recommendations is ensured by the personal commitments of its members. As the rate of implementation differs between countries, most often due to procedural differences, the WENRA secretariat monitors progress and reports annually at the WENRA plenary meetings. In principle, WENRA standards are not legally binding upon WENRA members. However, the fact that the WENRA members agree by consensus on all issued standards and recommendations ensures that they all do their best to effectively implement these requirements in their countries within a reasonable time.

10.3 The 'Waste Directive' of the European Union

The Council Directive 2011/70/EURATOM of 19 July 2011 (EU 2011), in short, the EU Waste Directive, is based on the Euratom Treaty and is legally binding for all Member States of the European Union. It is comprehensive and contains, among others, obligations concerning the national framework, the regulatory authority, the licence holders, financial resources, transparency, and reporting.

A quite fundamental requirement is that 'Radioactive waste shall be disposed of in the Member State in which it was generated.' However, deviations from this requirement are possible if a corresponding agreement exists between the Member State concerned and another Member State or a third country to use a disposal facility in one of them (EU 2011, Article 4.4). This means that multinational repositories, i.e. repositories that are located in one country but that would also accept the waste of some other countries, may be possible but are not the preferred option in the European Union. Such an option might be of interest to countries that have no waste from nuclear reactors but only waste from medicine, industry, and research.

In addition, the Waste Directive makes a clear distinction between high-level nuclear waste and low or intermediate-level radioactive waste: high-level radioactive waste, including spent nuclear fuel, shall be disposed of in deep geological repositories, while near-surface disposal is the typical disposal concept for low- and intermediate-level radioactive waste. This is because high-level waste has a considerably higher radiation level than low- and intermediate-level waste. For this reason, the Waste Directive implicitly considers it justified to apply shorter-term options, such as near-surface disposal, for low- and intermediate-level waste than for high-level waste. This is in accordance with international practice.

Another important aspect is the independence of the regulatory body. Article six of the Directive states that Member States must ensure that the competent regulatory authority is functionally separate from any other body concerned with the promotion or utilisation of nuclear energy or radioactive material, in order to ensure effective independence from undue influence on its regulatory function.

Member States have to submit a report on the implementation of the Directive to the European Commission every three years. The first such report was due in 2015. The Commission then reports to the European Parliament on progress made in the Member States on the implementation of the Directive, including an inventory of spent fuel and radioactive waste present in the Community's territory. In addition, every ten years, member states have to perform self-assessments on all aspects of the Directive's requirements and invite international peer reviews. The outcomes of such peer reviews have to be reported to the Commission and the other Member States.

Bibliography

EU 2011 *European Union, Council Directive* 2011/70/EURATOM, 19 July 2011 https://eur-lex.europa.eu/legal-content/EN/TXT/PDF/?uri=CELEX:32011L0070&from=FR

EU 2014 *European Union, Council Directive* 2014/87/EURATOM, 8 July 2014 https://eur-lex.europa.eu/legal-content/EN/TXT/PDF/?uri=CELEX:32014L0087&from=EN

IAEA 1989 *Statute* as amended up to 28 December 1989 https://iaea.org/sites/default/files/statute.pdf

IAEA 2006 *Fundamentals Safety Principles, Safety Fundamentals* SF-1 https://www-pub.iaea.org/MTCD/Publications/PDF/Pub1273_web.pdf

IAEA 2018 *Integrated Regulatory Review Service Guidelines* IAEA Service Series 37 https://www-pub.iaea.org/MTCD/Publications/PDF/SVS-37web.pdf

IAEA 2021a *List of Member States* as of 7 April 2021 https://iaea.org/about/governance/list-of-member-states

IAEA 2021b *Long Term Structure of the IAEA Safety Standards and Current Status* http://www-ns.iaea.org/committees/fileS/CSS/205/status.pdf [accessed 20 August 2021]

IAEA 2021c *Review Missions and Advisory Services* https://iaea.org/services/review-missions [accessed 1 September 2021]

IAEA 2021d *Joint Convention on the Safety of Spent Fuel Management and on the Safety of Radioactive Waste Management* https://iaea.org/topics/nuclear-safety-conventions/joint-convention-safety-spent-fuel-management-and-safety-radioactive-waste [accessed 1 September 2021]

WENRA 2006 *Harmonization of Reactor Safety in Member Countries, January 2006* https://wenra.eu/sites/default/files/publications/rhwg_harmonization__report_final.pdf

WENRA 2008 *Reactor Safety Reference Levels, January 2008* https://wenra.eu/sites/default/files/publications/list_of_reference_levels_january_2008.pdf

WENRA 2014a *WENRA Report, Waste and Spent Fuel Storage Safety Reference Levels, Version 2.2, April 2014* https://wenra.eu/sites/default/files/publications/wgwd_storage_report_final.pdf

WENRA 2014b *WENRA Report, WENRA Safety Reference Levels for Existing Reactors, Update in Relation to Lessons Learned from Tepco Fukushima Dai-ichi Accident, 24 September 2014* https://wenra.eu/sites/default/files/publications/wenra_safety_reference_level_for_existing_reactors_september_2014.pdf

WENRA 2014c *WENRA Report, Radioactive Waste Disposal Facilities Safety Reference Levels, 22 December 2014* https://wenra.eu/sites/default/files/publications/srl_disposal_final_version_2014_12_22.pdf

WENRA 2015 *WENRA Report, Decommissioning Safety Reference Levels, Version 2.2, 22 April 2015* https://wenra.eu/sites/default/files/publications/wgwd_report_decommissioning_srls_v2_2.pdf

WENRA 2017 *WENRA Umbrella Document, Interfaces and Interdependencies, Supplementary Report, August 2017* https://wenra.eu/sites/default/files/publications/umbrelladocument_sep_2017.pdf

WENRA 2018 *WENRA Report, Radioactive Waste Treatment and Conditioning Safety Reference Levels, April 2018* https://wenra.eu/sites/default/files/publications/report_radioactive_waste_treatment_and_conditioning_safety_reference_levels.pdf

WENRA 2021a *WENRA Report, WENRA Safety Reference Levels for Existing Reactors 2020, 17 February 2021* https://wenra.eu/sites/default/files/publications/wenra_safety_reference_level_for_existing_reactors_2020.pdf

WENRA 2021b *WENRA, Western European Nuclear Regulators' Association, WENRA Website, 01 September 2021* https://wenra.eu/

Chapter 11

Perspectives on risk and risk perception

Anne Eckhardt

Internationally, it is widely recognised that the management of radioactive waste is indispensable and permanent disposal solutions should be initiated without delay. Nevertheless, plans for disposal facilities and the transport of radioactive substances often meet with resistance, and dialogue with the interested public is challenging. This chapter therefore provides insights into the socio-scientific aspects of the management of radioactive waste, in particular, the perception of risks, safety, and uncertainty. Historical developments that may influence risk perception are covered in chapter 12 'The politicization and politics of nuclear waste: a socio-technical history.' For a better understanding of how to deal with safety, risk, and uncertainty, reference is made to part I, chapter 9 'Decision-making in the presence of uncertainty'. Safety Case as a tool supporting the development of deep geologic repository' in this book.

Safety is the main purpose of radioactive waste management. What 'safety' means exactly, however, is not obvious; neither are the level of safety that should be aimed for or how the different aspects of safety are sometimes to be balanced against each other. Safety and risk, especially radiological risks, lead to controversies about the management of radioactive waste. Differing positions are widespread. While, in the following two quotations, Greenpeace declares that the risks are unacceptable due to a lack of fairness, the World Nuclear Association WNA cites good regulation and safety measures and states that there is no risk at all:

> "All the way through the nuclear chain, local populations are subjected to increased health risks, and yet more often than not they have not been asked if they are willing to put up with those increased risks (Greenpeace 2019, p 20). International conventions define what is hazardous in terms of radiation dose, and national regulations limit allowable doses accordingly. Well-developed industry technology ensures that these regulations are met so that any hazardous waste is handled in a way it poses no risk to human health or the environment (WNA 2020)."

doi:10.1088/978-0-7503-3095-4ch11

There may be political and tactical reasons behind many controversial statements on radioactive waste management. However, these are not the only reasons for controversies. Often, conflicts over the disposal of radioactive waste can be traced back to different perceptions of risk, which manifest themselves in opinions and commitments in favor of certain solutions or against them.

For everyone interested in good and successful radioactive waste management, it is therefore useful to understand how different actors perceive and assess risks. The risks associated with radioactive waste have some specific characteristics that act upon risk perception. Together with more general insights into risk perception, these specific characteristics should be considered when communicating about risks and progressing toward acceptable and accepted solutions in radioactive waste management.

11.1 Safety, risk, and uncertainty

In radioactive waste management, safety is generally considered to be the main objective and is more important than other aims, such as fairness, justice, or economic efficiency (see part II, chapter 14 in this book). The term safety, as used here, also includes security. Nuclear security refers to 'the prevention and detection of, and response to, criminal or intentional unauthorized acts involving nuclear material, other radioactive material, associated facilities or associated activities' (IAEA 2018, p 155).

Over the course of time, it became clear that people have different understandings of the concept of safety. Whereas some understand safety as the absence of intolerable risks, others only speak of safety when there is no risk at all. In a given situation, some understand safety as being well protected by trustworthy institutions and persons, but others understand it to mean that they can control the risks by themselves. Some accept safety easily, while others demand precise guarantees and adherence to certain moral norms to consider a given situation as safe. The specialists from the various scientific and technical disciplines involved in radioactive waste management use different concepts of safety, and different concepts are also widespread among other stakeholders.

Because of the manifold aspects that can be associated with safety, the Forum on Stakeholder Confidence FSC of the Nuclear Energy Agency NEA states that safety is a 'multi-dimensional concept.' Technical safety, for instance, is supplemented by security aspects, fair decision-making processes are gaining importance in safety issues, and as a social construct, safety also has an emotional component (NEA 2013, p 39). Moreover, assigning the attribute 'safe' is a normative act based on the criteria for what is considered safe or 'safe enough.'

The International Atomic Energy Agency IAEA sets a safety objective for all circumstances that give rise to radiation risks, namely, 'to protect people and the environment from harmful effects of ionizing radiation' (IAEA 2006, p 3f). Complete protection, however, is impossible. Therefore, the IAEA understands safety as a state in which the remaining risks are acceptable.

Various notions exist not only for safety but also for risk. 'Risk' can be associated with a lack of information, deviations from planned objectives, or the willingness to take responsibility in the face of the possibility of damage—just to mention a few examples. In the context of technology and technical installations, risk is often understood as the possibility of damage. Damage occurs if something that humans value is impaired. In accordance with the protection objective of the IAEA, in radioactive waste management, the damages primarily considered are those that affect people and the environment.

Risks can be assessed and calculated, usually by using information about the probability of occurrence and the extent of damage. With regard to the long-term disposal of radioactive waste, risk is often evaluated using the probability and the consequences of scenarios of future development. If there is too little information available to assess a risk, uncertainty prevails. Uncertainties are particularly significant in the case of the final disposal of high-level radioactive waste. Here, important sources of uncertainty are the long assessment periods over which the safety of disposal is to be demonstrated, the complexity of the sociotechnical disposal system, and spatial heterogeneities in the natural environment of the repository. A deep geological repository is a unique specimen, and its long-term operation can hardly be based on the experience gained with similar facilities, as is usually the case in engineering.

A pragmatic typology of uncertainties relies on the availability and the use of information cf. figure 11.1.

Radioactive waste management involves a variety of risks. The waste itself poses radiological risks and risks due to chemotoxicity. Risks may also arise due to criticality, heat generation, or chemical reactivity. Moreover, waste management activities are linked to a number of additional risks. During conditioning and

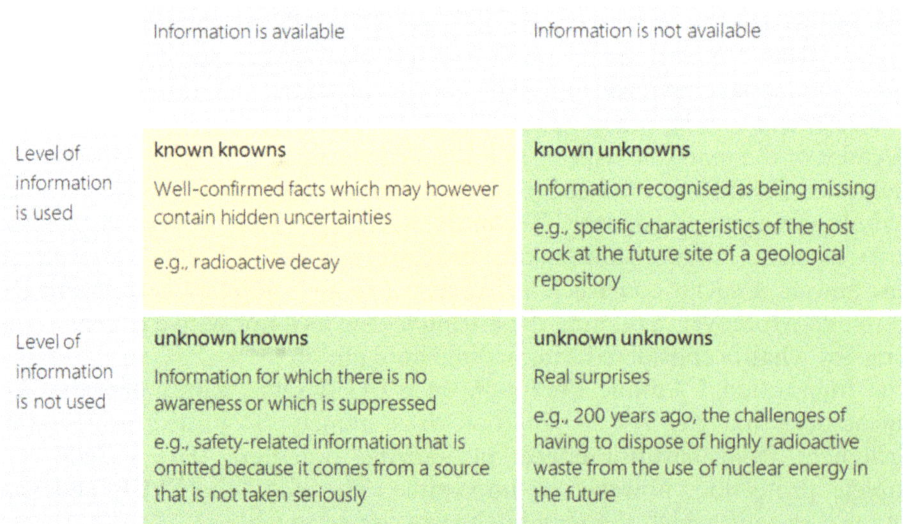

Figure 11.1. A pragmatic typology of uncertainties in the management of radioactive waste (Eckhardt 2016).

transportation as well as in the course of the construction and operation of disposal facilities, accidents can occur in which people are injured and property damage is caused. Heat, dust, and natural radiological emissions during the construction and operation of deep geological repositories may impair workers' health. Psychological stress might affect persons involved in disposal activities or resisting them, and so forth. In the national regulations, different types of risk are typically covered by separate rules and standards. A holistic view of the risks associated with radioactive waste management is challenging and rarely provided.

Such a holistic view raises questions about risk assessment that cannot be easily resolved. How should, for example, risks that occur in the near future be weighed against risks that may occur in the distant future? This affects, for instance, the monitoring of a deep geological repository. People are exposed to risks when installing and maintaining underground monitoring equipment, and perhaps also when viewing monitoring devices. These risks, which are well known today, have to be traded off against an uncertain safety gain in the future, which is due to the fact that monitoring can reduce uncertainties about the long-term safety of a repository.

Similar questions are: how are individual and collective risks to be weighed against each other? Can and shall an increased risk for individuals or smaller groups of people be accepted in order to reduce the overall risk for all? How should different types of damage be compared to each other, for example adverse health effects on people versus adverse changes to ecosystems? How should we deal with damage that is indirect and/or difficult to quantify, for example, psychological impairment due to conflicts over a storage site?

These questions must essentially be reflected upon from ethical (see part II, chapter 14 in this book) and legal (see part II, chapter 10 in this book) perspectives.

11.2 'Creeping danger' and 'imminent threat'

Among the specific risks in the management of radioactive waste, compared to the management of most other types of waste, are the risks associated with low doses of ionising radiation during transportation and the operation of processing, storage, and disposal facilities. They represent a type of risk that has been described by the sociologist O. Renn as a 'creeping danger.' Creeping dangers are characterized by a long delay between exposure and effect, no possibility of detecting the danger using the human senses, and dependence on third parties to provide information about the risk. People affected by creeping dangers usually cannot verify whether the (differing) expert opinions about the safe management of radioactive waste are true. If a person develops cancer in the vicinity of a storage or disposal facility, it is not clear whether ionising radiation from the waste is the cause or whether there is another relevant reason, for example, a natural influence or an accidental event in the body of this person. In general, humans prefer a meaningful explanation for an event to the insight that this event has happened by chance. People tend to identify themselves with victims and to look for culprits who are responsible for their suffering. This contributes to social resistance against storage and disposal facilities for radioactive waste (Streffer *et al* 2011, p 340f).

A second type of risk which is relevant for radioactive waste management from a social science perspective is the 'imminent threat.' Imminent threats are risks that combine a low probability of occurrence with a high level of damage. This type of risk plays a major role in debates on nuclear power plants. In the case of radioactive waste management, its significance is somewhat lower, but still present. For persons who judge certain damages, especially catastrophic damages, as unacceptable in principle, the associated imminent threats are also intolerable. Such persons may therefore categorically reject processing, storage and disposal facilities or the transport of radioactive waste—irrespective of the safety measures taken (Streffer *et al* 2011, p 338f). Their attitude towards imminent threats might be supported by the fact that many persons have fundamental difficulties in dealing with events whose occurrence is uncertain. 'Statistical illiteracy' has been well studied in different countries, aspects of life, and professional environments, especially in the medical field, and even affects professionals who deal with risks (Gigerenzer *et al* 2007, Gigerenzer 2010). Some of its impact on risk perception, however, may be less related to a lack of mathematical skills than to the fact that the corresponding information does not emotionally appeal to the recipients (Chauvin 2018, p 53; see Individual risk perception).

11.3 Controversies about risk and uncertainty

In society, some risks lead to controversies due to ethical and moral issues. Synthetic biology raises the question of the extent to which humans are allowed to shape nature, design new creatures, and 'play God.' Bioelectronics is connected to questions concerning the boundaries between man and machine and the societal understanding of humanity. The disposal of radioactive waste poses controversial questions, such as what a fair site selection procedure for a hazardous facility looks like, whether compensation is justified or rather undermines a fair site selection procedure, how future generations can best be protected from the harmful effects of radioactive waste, how—in the sense of sustainable disposal—their freedom of action is best guaranteed, and whether precautions are required to permanently remind mankind of the problems of nuclear energy (see part II, chapter 14 in this book). Where such 'ambiguous risks' prevail, risk perception, moral considerations, and affects are closely intertwined.

In dealing with ambiguous risks, the International Risk Governance Council (IRGC) recommends a flexible, adaptable approach that can take account of changes in societal conditions. All groups that have special interests or value commitments with respect to the risk or the benefits associated with the risk should be involved in discourse-based decision-making. This aims to create tolerance and mutual understanding of conflicting views and values, build trust, and ultimately find solutions that are acceptable to all concerned (IRGC 2017, p 18f). Adaptable, participatory approaches to radioactive waste management have already been implemented in some countries. In Canada, an adaptive site selection process was developed collaboratively with Canadians and Aboriginal people. This process proceeds with interested communities only. In Germany, great efforts have been made in recent years to launch a new site selection procedure (see part II, chapter 16 in this book). This procedure is based on a learning, self-questioning process, which provides opportunities for the interested public to have their say. Other countries which have implemented adaptable participatory site selection processes successfully are Sweden and Switzerland.

Risks are quantifiable or can at least be assessed qualitatively. Therefore, whether a risk is acceptable or not can frequently be decided with the help of quantitative, semi-quantitative, or qualitative thresholds. Uncertainties are sometimes quantifiable, but in many cases, they are not, especially when unknown unknowns or unknown knowns come into play. It is often difficult to judge whether an uncertainty is acceptable.

Some information relevant to the risk assessment of waste management facilities such as geological repositories can be empirically proven. But there exists no proof that the expected or deviating scenarios considered in safety assessments will cover or envelop the future evolution of the repository system which will actually occur. Moreover, different interests and value attitudes shape different constructions of future reality. Therefore, uncertainties about the safety of waste management facilities often become the subject of controversy and can also be used tactically to question waste management projects.

Uncertainties provide suitable arguments both for and against waste management projects. A fundamental argument against waste management projects is that the remaining uncertainties are still or principally too great (see e.g. Shrader-Frechette 1993). An argument in favour of such projects is that they reduce safety-relevant uncertainties, for example, because a new facility—whether it is a better designed and equipped interim storage facility or a final repository in deep geological formations—offers more robust protection against hazards.

Uncertainties are especially relevant for the long-term project of the final disposal of high-level radioactive waste. Final repositories in the deep geological underground were explicitly chosen to minimise the influence of human activities on the safety of the repository. Developments such as climate change, energy system transformation, and technological progress, however, beg the question of whether the safety of a final repository can actually be provided for hundreds and thousands of years. So far, if the long-term safety of a repository has to be demonstrated, it is generally shown that a repository system largely undisturbed by humans meets the safety requirements during the assessment period. Human activities currently only play a subordinate role in this respect. In the future, anthropogenic developments will probably have to play a more important role in demonstrating safety.

From a governance perspective, the recommendation for addressing uncertainties is precaution-based and resilience-focused management which is open to revising decisions and coping with surprises (IRGC 2017, p 25).

When considering societal views of risk, however, it must not be disregarded that radioactive waste management as a whole provides safety. This benefit is actually recognised by the public and weighed against risks and uncertainties. A broad representative survey in the European Union, for instance, revealed an overwhelming consensus that a solution for managing high-level radioactive waste should be found now, rather than leaving it for future generations. However, a clear majority of the respondents believed that there was no safe way to dispose of this waste, and less than half of them considered deep geologic disposal to be the most appropriate solution for the long-term management of high-level radioactive waste (European Commission 2008, p 91).

11.4 How individual persons perceive risk

In everyday life, different notions of risk, uncertainty, and safety are common. Some people welcome vaccinations because they reduce the risk of getting a serious infection. Others are reluctant, due to the risks associated with the vaccination itself. Some people hesitate to book a flight because they are concerned about the risk of aviation crashes. Others tell them that it is more dangerous to travel the same distance by car than by plane. Research in various disciplines, such as anthropology, behavioural economics, differential psychology, evolutionary biology, neuroscience, and social psychology contributes to the understanding of different notions of risk and the factors which influence them.

Intensive research on the ways in which different groups of people quantitatively perceive risk started in the 1980s. The psychologist Paul Slovic compared the differing risk perceptions of experts and 'laypersons'. He identified two major determinants of risk perception among laypersons: the degree to which a risk is dreaded and the degree to which a risk is unknown (Finkel 2008, p 123). In both cases, radioactive waste management is in an unfavourable position: the risks associated with radioactive waste are dreaded because they are difficult to control and they have the potential to be catastrophic if large quantities of radionuclides are released, especially if they are released over a short period of time. Radioactivity cannot be perceived by the human senses. The disposal of radioactive waste requires specialised knowledge, and many people are unfamiliar with how to measure radioactivity and how to protect themselves from radioactivity. Another unknown is the way in which radioactive waste management will develop over longer periods of time—especially in view of unknown future societal and technological developments that might influence the risks associated with the long-term disposal project (figure 11.2).

What distinguishes experts in a particular risk from members of the public looking at the same risk is mainly a different background of knowledge and experience. The experts' perception and handling of the risk is influenced by their special level of information and experience, professional rules, and standards, as well as by specific convictions within their professional community. Psychological research has shown that the risk perception of experts correlates better with annual fatalities than the perception of the public. Members of the public, however, have a broader conception of risk. Both types of risk perception, that of experts and that of the public, follow their own rationality. Both have a systematic nature and are predictable, at least to a certain extent (Chauvin 2018, p 53).

Therefore, it is ineffective to assume that a risk only needs to be explained better to the public so that people's 'subjective' perception of the risk will align with the

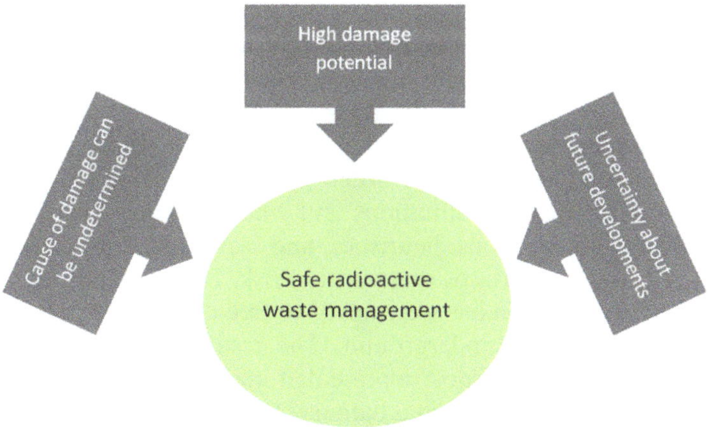

Figure 11.2. Fundamental aspects which cast doubt on the safety of radioactive waste management from a sociological perspective (own illustration, based on Streffer *et al* 2011).

experts' 'objective' perception of the risk. However knowledge and information about the causes, characteristics, and consequences of a risk can indeed have an impact on risk perception. A representative survey in Switzerland on the topic of carbon dioxide capture and storage, for instance, showed that many people have only limited information that allows them to weigh up risks and benefits against each other. In their risk assessment, intuitive concepts related to underground carbon dioxide storage play an important role. Such intuitive concepts often deviate significantly from the characteristics of projects which are actually considered for carbon dioxide capture and storage. The risks associated with these concepts are easily affected by contextual factors and unstable. A challenge for communication is therefore to provide information that contributes to a better understanding using the short attention span people are willing to devote to this topic (Wallquist *et al* 2010).

In general, psychological research shows that more knowledge tends to contribute to a better acceptance of risks that do not evoke moral concerns, whereas it can impair the acceptance of risks which are morally controversial (Visschers and Siegrist 2018, p 67).

This insight refers to the importance of attitudes, values, and worldviews in the perception of risks associated with radioactive waste management. Several studies have revealed correlations between value orientations, such as communitarian, egalitarian, hierarchical, and individualistic orientations and risk perception (Visschers and Siegrist 2018, p 72, Siegrist and Árvai 2020, p 5). Research conducted in the United States, Japan, and the Netherlands suggests relationships between worldviews and the risk perception of nuclear energy. Altruism and biospherism, meaning a high concern for nature and the environment, tend to be consistently negatively correlated with the perceived risks of nuclear energy use (Siegrist and Árvai 2020, p 5), whereas people who value hierarchy, tradition, and individualism rather support the use of nuclear power for energy production (Visschers and Siegrist 2018, p 72). A key topic in the context of radioactive waste management is fairness, in particular, procedural fairness when it comes to the selection of sites for waste management facilities. The acceptance of a hazard can be increased by a fair decision procedure, even if people are concerned about its risks (Visschers and Siegrist 2018, p 74).

Human evolution has led to 'slow' and 'fast' thinking (Kahnemann 2011); feelings and rationality interact effectively in order to arrive at helpful decisions, even in complex and incomprehensible situations and when it comes to dealing with risk. This is why affects and emotions, heuristics, and biases also have their place in the perception of risk and in decisions that concern risk. In an early study conducted in the USA, more than 3000 persons were asked to free-associate with a nuclear waste repository in the geological underground. The results of this investigation were overwhelmingly negative. The most represented image categories were 'dangerous/toxic,' 'death/sickness,' 'environmental damage' for the consequences of a repository and 'bad/negative,' 'scary,' 'unnecessary/opposed,' 'not near me,' and 'war/annihilation' associated with the concept of a repository (Slovic *et al* 1991, p 1605). Among other things, the perception of radioactive waste management is thus apparently linked to perceptions of nuclear energy use and nuclear weapons.

Psychological research has shown that the assessment of risk is influenced by affects, for example the feelings that a risk elicits (Visschers and Siegrist 2018, p 69f). Here, the management of radioactive waste is obviously in a weak position, because it arouses negative feelings. In some countries and contexts, these negative feelings may be related, amongst others, to bad experiences with the management of radioactive waste in the past. Heuristics, simple mental strategies for making rapid decisions in complex situations, and biases, which give certain cognitive aspects a disproportionate weight, can also have an impact on risk perception. Heuristics and biases affect both the experts' and the public's risk perceptions and assessments. Experts are particularly affected when they have to go beyond the limits of available scientifically substantiated information and rely on intuition (Chauvin 2018, p 53f). Currently, there is still a need for a better understanding of the factors that determine which heuristics people rely on when evaluating hazards (Siegrist 2019).

The perception of risk varies greatly between individuals (figure 11.3). Based on the current state of research worldwide, several possible influences underlie individual variations in the perception of the risks associated with radioactive waste management (Lauriola and Weller 2018, p 3f, Chauvin 2018, p 37f). In essence, these can be traced back to differences in personality, individual experiences, value orientations, worldviews, and the social and cultural environment of a person. Gender and ethnicity also have a confirmed influence on risk perception. Persons who are female or non-white tend to perceive risks as more threatening than white men. This tendency, backed by numerous studies, can be explained by the effect, that 'those who have power and control over risks, who benefit from them, who trust in authorities and experts, and who do not see themselves as vulnerable or as a target of environmental injustice tend to perceive risks as low compared to others' (Chauvin 2018, p 45). The influence of gender and ethnicity thus at least overlaps with that of the social and cultural framework.

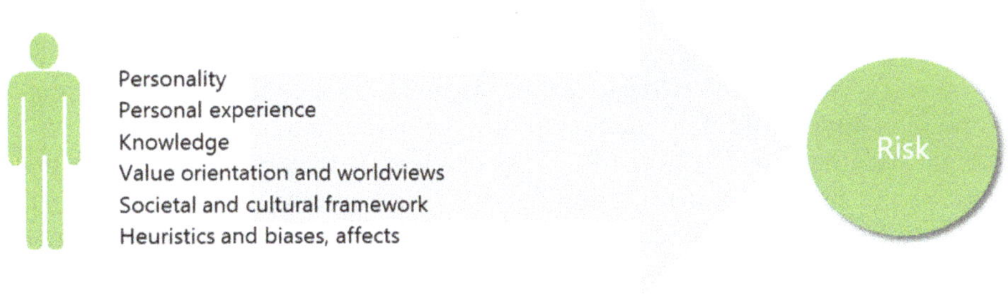

Figure 11.3. Influences on the individual perception of risks associated, for instance, with radioactive waste management (own illustration, based on the previously cited literature).

Research into risk perception does not always clearly distinguish between risks and uncertainties. As a result, some studies of risk perception also contain information on the perception of uncertainties. Specific research into the perception of uncertainties has increasingly been carried out in recent years, particularly in the context of climate change and health-related decisions. This research suggests that uncertainties referring to missing data or those related to a lack of precision in results are more likely to be accepted than if data are contradictory or experts dissent (Broomell and Kane 2017, p 288).

11.5 Risk perception, trust, and international divergence

Trust is a psychological state. Those who trust others have positive expectations of their intentions or behaviour. Trust implies an acceptance of vulnerability because it implies reliance on others (Rousseau *et al* 1998, p 395). In a complex society which involves a high degree of division of labour, trust is indispensable. Not trusting others is often impossible, because the controls that would be required in this case exceed the available resources (Siegrist 2019).

When projects for the management of radioactive waste are debated, the discussion often leads to trust: a lack of trust between the nuclear industry, authorities, and the public; trust that was betrayed in the past; and trust that the regulators and implementers want to regain. In many countries, this situation has hardly changed over recent decades:

> "The Department of Energy's program for disposing of high-level radioactive wastes has been impeded by overwhelming political opposition fueled by public perceptions of risk […] the mishandling of radioactive wastes from the nation's military weapons facilities has contributed toward creating a profound state of distrust that cannot be erased quickly or easily (Slovic *et al* 1991, p 1603). From decades of secrecy, bureaucracy, and top-down decision-making, Americans distrust those who control nuclear waste decisions. In the absence of community engagement, Americans don't recognize the acute need for nuclear waste storage, and perceive such storage as fallible and dangerous (Jennewein 2018)."

Accordingly, the relationship between trust and risk perception has attracted scientific interest, and numerous studies have been devoted to it (Siegrist 2019). Research so far suggests that trust can be strongly correlated with risk perception, when people have little knowledge about an issue that is important to them (Visschers and Siegrist 2018, p 69). This applies, for example, to most persons affected by the site decision for a radioactive waste management facility. Yet it is difficult to differentiate between trust and other influences on risk perception. Trust is linked to affects about risks as well as to values and worldviews (Siegrist 2019); common value orientations strengthen the trust between different stakeholders (Visschers and Siegrist 2018, p 75). The personality of the risk perceiver can be of importance. Optimistic people probably tend both to perceive risks as less threatening and to trust other people and institutions more than pessimists. It is difficult to clarify which of these influences actually shapes risk perception in a specific situation (Siegrist 2019, p 8).

As some influences on individual risk perception seem to be deeply rooted in human psychology, it is plausible that aspects of risk perception can claim universal validity. Indeed, the results of early psychometric research into risk perception have been confirmed by studies in many different countries and cultures (Siegrist and Árvai 2020, p 2). The risk perception of individual persons is, however, and as mentioned before, influenced by patterns of value orientations and worldviews. These patterns can differ between cultures. A culture's position on the individualism–collectivism continuum, for instance, has an impact on the risk perception of risky options. As such, based on identical information a risk is perceived as greater by people in the US than by those in China (Siegrist and Árvai 2020, p 7).

In the response to the risks associated with radioactive waste management, clear disparities can be observed between different countries: while the transport of high-level radioactive waste led to fierce protests in Germany, they hardly attracted any attention in neighbouring Switzerland. The majority of the Finnish population shows a pragmatic attitude in dealing with the final disposal of radioactive waste, whereas in France—which, like Finland, is committed to the continued use of nuclear energy—controversies are more pronounced. Historical, political, and cultural influences contribute to such country-specific differences. Recent research shows that in Finland, a high level of trust in institutions, in particular, in the safety authority Säteilyturvakeskus (STUK), concurs with a high level of trust in the state. This favours consensual solutions for the management of radioactive waste. In France and Germany, by contrast, distrust in state institutions is deeply rooted when it comes to radioactive waste management, due to historical developments (Lehtonen 2020).

Switzerland has so far successfully pursued a site selection procedure for a deep geological repository that includes a wide range of actors in consultation. This procedure is in line with its specific political culture and its combination of representative and direct democracy, and is therefore not transferable to other countries without further consideration (Hocke and Kuppler 2015, p 157f).

11.6 Dealing with different risk perceptions in practice

Some lessons can be learned from current knowledge about the perception of safety, risk, and uncertainty in order to address differing perspectives on radioactive waste management (see also NEA 2017, IGSC 2020):

- Do not talk in general terms about safety, risk, or uncertainty, but take a more differentiated approach to the topic under discussion.
- Try to get to the bottom of the causes of possible dissent and to understand the concerns of the other side well before arguments for or against a position are put forward.
- Take into account the diversity of individual risk perceptions. You will never enter into dialogue with a homogeneous group, but will always encounter a variety of individual risk perceptions.
- In dealing with risks, everyone is both a layperson and an expert. Keep in mind that the risk perception of experts and laypersons alike follows a rationality, even though these rationalities may differ. Treat all interlocutors with full respect.
- Common value orientations strengthen trust and facilitate dialogue on risk. Therefore, look for a common value basis on which you can talk about risk.
- Consider that risk perception and the tactical use of arguments on risk, security, and uncertainty are often intertwined in discussions.
- If you represent the side of the regulator or the implementer, address uncertainties about safety-relevant topics openly, but at the same time prove competence in dealing with these topics.

11.7 Shaping the 'risk profile' in dialogue

As we have seen, research so far casts a rather dark light on the perception of the risks in radioactive waste management:

Risks associated with radioactive waste management are dreaded by the public because they include a great damage potential—largely regardless of the probability of occurrence. Ionising radiation can cause creeping diseases such as cancer, which reinforces the perception of the risks as menacing. When cases of cancer occur in the vicinity of a waste management facility, they are often attributed to the radioactive waste, even if the real causes are most likely other ones, for example the exposure to natural radiation or accidental events in the human body. Radioactivity cannot be perceived with the human senses and is therefore difficult for members of the public to control. The disposal of radioactive waste requires specialised knowledge, and many people are not very familiar with how to measure radioactivity and how to protect themselves from radioactivity. The resulting lack of controllability intensifies the perception of risk as threatening.

Trust is therefore essential, especially trust between the representatives of the regulator and implementer and the public. However, trust between stakeholders in radioactive waste management is currently impaired in many countries. This is mainly due to negative experiences that stakeholders have had with each other in the past (see, as an example, the highlighted quotes from Slovic and Jennewein in chapter 2.5). These unfavourable experiences probably also contribute to the fact that radioactive waste management triggers predominantly dreary associations. Accordingly, the risk perception of many persons is characterised by negative feelings and emotions when referring to radioactive waste.

The risks associated with radioactive waste are balanced by one key benefit—the reduction of the risks arising from the waste through appropriate waste management and, in particular, deep geological disposal. When it comes to distributing the risks of disposal, procedural fairness in the site selection processes for waste management facilities is of special importance. Humans have a deep-rooted sensibility to fairness (Finkel 2008, p 130). If diverse stakeholders are involved in processes for the management of radioactive waste, it must be taken into account that different values then encounter each other, which may lead to different risk perceptions—apart from the differing political and economic interests which also often occur.

Fortunately, most of these insights about risk perception are far from new. International committees, such as the FSC of the NEA, are concerned with the broad spectrum of risk perceptions and issue recommendations on how to deal constructively with it. An increasing number of countries are involving stakeholders in the management of radioactive waste, especially in the context of site selection procedures and are gathering positive experiences with it (cf. part II, chapter 13 'Role of stakeholder involvement in the implementation of radioactive waste management projects'). Even in research, efforts are made to evaluate the integration of different risk perceptions in safety assessments for the final disposal of radioactive waste (Röhlig *et al* 2021).

Radioactive waste management is not the only area in which widely differing risk perceptions arise. In other areas with controversial risks, such as climate change, mobile communications, genetic engineering, or digital transformation, a dialogue between stakeholders with different risk perceptions is also required. For this reason, more and more experience is being built up that describes how to achieve a constructive and solution-oriented dialogue between people with different risk perceptions. Activities in related areas, such as carbon capture and storage, can contribute to a broader view of the risks associated with radioactive waste management by allowing both sectors to benefit from mutual experience and newly gained knowledge.

In the future, the understanding of different risk perceptions is likely to improve further through new research. It can hopefully be expected that initiatives worldwide will progressively contribute to a constructive handling of different risk perceptions in the field of radioactive waste management. Many regulators and implementers have learned to communicate about risks and uncertainties in a more differentiated and open way than they did before. Young people, who bring new perspectives and innovative ideas and are less burdened by past conflicts, contribute and take the lead. Perhaps they will also find new starting points that will raise interest in this topic, such as the challenge of increasing the sociotechnical resilience and robustness of geological disposal or preserving knowledge about risk over very long periods of time.

"Following several decades of committed implementation of disposal strategies in Finland and Sweden, as well as cooperation in the development of a safe disposal solution based on a Swedish design, the first ever deep geological repository for spent fuel is being constructed in Olkiluoto, Finland (Mikhailova 2019). Radioactive waste will, in some cases, remain hazardous to humans for tens of thousands of years. How do we warn future generations 10 000 years from now not to go near a certain area? Well… […] We like the Ray Cats. Have you got any ideas for how to keep a warning alive for 10 000 years? (Fabbri 2018)."

Successful projects and new scientific perspectives open up different views of the risks associated with the disposal of radioactive waste. Thus, the perception of risks will continue to develop and change in the future.

Bibliography

Broomell S P and Kane P B 2017 Public perception and communication of scientific uncertainty *J. Exp. Psychol: Gen 2017* **146** 286–304

Chauvin B 2018 Individual differences in the judgment of risks: sociodemographic characteristics, cultural orientation, and level of expertise *Psychological Perspectives on Risk and Risk Analysis. Theory, Models, and Applications* ed M Raue, E Lermer and B Streicher (Cham: Springer)

Eckhardt A 2016 Comparative assessment of uncertainties and risks associated with different disposal options for radioactive waste *Proving Futures and Governing Uncertainties in Technosciences and Megaprojects. Foreknowledge Assessment (Proc. Paris Conf. 2016)* http://www.foreknowledge2016.com/doc/Andra-Foreknowledge_Proceedings.pdf [accessed 3 January 2022]

European Commission 2008 *Attitudes Towards Radioactive Waste* Special Eurobarometer 297

Fabbri P 2018 *The Ray Cat Solution* https://paolofabbri.it/recensioni-e-commenti/the-ray-cat-solution/ [accessed 3 December 2020]

Finkel A M 2008 Perceiving others' perceptions of risk. Still a task for Sisyphus *Ann. N. Y. Acad. Sci.* **1125** 121–37

Gigerenzer G 2010 Collective statistical illiteracy. Invited commentary *Arch. Intern. Med.* **170** 468–9

Gigerenzer G, Gaissmaier W, Kurz-Milcke E, Schwartz L M and Woloshin S 2007 Helping doctors and patients to make sense of health statistics *Psychol. Sci. Public Interest.* **8** 53–96

Greenpeace 2019 *The Global Crisis of Nuclear Waste* (Paris: Greenpeace France)

Hocke P and Kuppler S 2015 Participation under tricky conditions. The Swiss nuclear waste strategy based on the sectoral plan *Nuclear Waste Governance* ed A Brunnengräber, M R Di Nucci, A M Isidoro Losada, L Mez and M A Schreuers (Wiesbaden: Springer)

IAEA (International Atomic Energy Agency) 2018 *Safety Glossary. Terminology Used in Nuclear Safety and Radiation Protection* 2018 edn (Vienna: IAEA)

IAEA (International Atomic Energy Agency) 2006 *Fundamental Safety Principles* (Vienna: IAEA)

IGSC (Integration Group for the Safety Case) 2020 *Two Decades of Safety Case Development* An IGSC 20th anniversary brochure (Paris: IGSC)

IRGC (International Risk Governance Center) 2017 *Introduction to the IRGC Risk Governance Framework. Revised Version* (Lausanne: EPFL IRGC)

Jennewein M 2018 *Looking for a Trash Can: Nuclear Waste Management in the United States* Blog Science Policy (Harvard University) http://sitn.hms.harvard.edu/flash/2018/looking-trash-can-nuclear-waste-management-united-states/ [accessed 3 December 2020]

Kahneman D 2011 *Thinking, Fast and Slow* (New York: Macmillan)

Lauriola M and Weller J 2018 Personality and risk: beyond daredevils—risk taking from a temperament perspective *Psychological Perspectives on Risk and Risk Analysis. Theory, Models, and Applications* ed M Raue, E Lermer and B Streicher (Cham: Springer)

Lehtonen M 2020 *Nuclear Power, Ideological Trust, (de)Politization and Mistrustful Civic Vigilance—Finnish, French, German and Spanish Experiences* http://berlin-ec.com/salzburg2020/files/Lehtonen.pdf [accessed 18 November 2020]

Mikhailova N 2019 *Developing the First Ever Facility for the Safe Disposal of Spent Fuel* IAEA News https://iaea.org/newscenter/news/developing-the-first-ever-facility-for-the-safe-disposal-of-spent-fuel [accessed 3 December 2020]

NEA (Nuclear Energy Agency) 2017 *Communication on the Safety Case for a Deep Geological Repository* No. 7336 (Paris: NEA)

NEA (Nuclear Energy Agency) 2013 *Stakeholder Confidence in Radioactive Waste Management. An Annotated Glossary of Key Terms* NEA No. 6988 (Paris: NEA)

Röhlig K J, Ebeling M, Eckhardt A, Hocke P and Krütli P 2021 Transdisciplinary research on repository safety: challenges and opportunities. Conference abstract. *Saf. Nucl. Waste Disposal* **1** 205–7

Rousseau D M, Sitkin S B, Burt R S and Camerer C 1998 Not so different after all: a cross-discipline view of trust *Acad. Manag. Rev.* **23** 393–404

Shrader-Frechette K S 1993 *Burying Uncertainty. Risk and the Case Against Geological Disposal of Nuclear Waste* (Oakland CA: University of California Press)

Siegrist M and Árvai J 2020 Risk perception: reflections on 40 years of research *Risk Anal.* **40** 2191–206

Siegrist M 2019 Trust and risk perception: a critical review of the literature *Risk Anal.* **41** 480–90

Slovic P, Flynn J H and Layman M 1991 Perceived risk, trust, and the politics of nuclear waste *Science* **254** 1603–7

Streffer C, Gethmann C F, Kamp G, Kröger W, Rehbinder E, Renn O and Röhlig K J 2011 Radioactive waste. Technical and normative aspects of its disposal *Ethics of Science and Technology Assessment* vol 38 (Berlin: Springer)

Visschers V H M and Siegrist M 2018 Differences in risk perception between hazards and between individuals *Psychological Perspectives on Risk and Risk Analysis. Theory, Models, and Applications* ed M Raue, E Lermer and B Streicher (Cham: Springer)

Wallquist L, Visschers V H M and Siegrist M 2010 Impact of knowledge and misconceptions on benefit and risk perception of CCS *Environ. Sci. Technol.* **2010** 6557–62

WNA (World Nuclear Association) 2020 *Radioactive Waste—Myths and Realities* (Updated February 2020) https://world-nuclear.org/information-library/nuclear-fuel-cycle/nuclear-wastes/radioactive-wastes-myths-and-realities.aspx

IOP Publishing

Nuclear Waste
Management, disposal and governance
Klaus-Jürgen Röhlig

Chapter 12

The politization and politics of nuclear waste: a socio-technical history

Karena Kalmbach

12.1 Introduction[1]

Around the globe, diverse endeavours are under way to identify disposal sites or to build repository facilities for high-level nuclear waste. These projects are at different stages of the planning and building process. However, at the time of writing, not a single long-term disposal or repository facility for high-level nuclear waste is in operation. However, there are numerous projects that have been cancelled—again, at all the different stages of the planning or building process. Among these projects, Yucca Mountain in the USA and Gorleben in Germany are probably among those best known and referred to in public debates. What is more, many projects for high-level waste disposal or long-term repository sites have been severely delayed, or have never even started. For other projects, it is unclear what will happen once the building stage of the facility is indeed finalized: what types and amounts of waste will be disposed of or stored there? How will the waste be packaged? Will the disposal process go all its way to the final closure of the site, or will plans be reversed or waste even be retrieved?

It is often claimed that without politization—politization of either the site, the search process, the legal licensing procedure, or the classification of the waste itself—we would be already much further ahead with the challenge of safely handling or disposing of nuclear waste. According to these claims, there have been concrete scientific solutions at hand, but then politics kicked in—and that is when problems started. Other actors turn this argument upside down. They claim that the reason that there is no disposal facility already in operation and the reason that current projects involve so many uncertainties lie in the fact that previous decisions about

[1] I am grateful to Klaus-Jürgen Röhlig, Armin Grunwald, Jan-Henrik Meyer, and Jochen Ahlswede for their helpful comments on earlier versions of this chapter.

sites, search processes, licensing procedures, and waste classifications have been primarily political: from the very start, politics was given priority over scientific facts—and then problems occurred because of the political subtext to the allegedly 'neutral' facts.

Although these arguments stand in opposition to each other, they interestingly unite in the solution they advocate: abandoning politics and focussing on scientific truth and nothing but scientific truth. However, science is inherently social and thus also political. There is no universal answer to the questions of which kind of scientific research society should foster or which risks linked to scientific research society has to accept. These questions are negotiated in specific contexts, and the answers change over time and place. Thus, it is barely surprising that politics and politization have played a role in nuclear waste management. It is rather surprising that this is continuously presented as an astonishing revelation—and that the claim for 'neutral' science seemingly always increases despite the obvious normativity of nuclear debates.

Thus, this chapter does not present a selection of examples in which politization in nuclear waste management projects took place, nor does it unveil occasions on which decision-making has been primarily political. There are shelves full of books written on these topics, and other authors have already made great efforts to synthesise these various findings (see, for instance, the recent publications which came out of the research project 'History of Nuclear Energy and Society' (HoNESt 2020)). Instead of adding yet another article to this convolute, this text will take the next level up and look at the debates about 'politization' and 'politics' from a meta-perspective, drawing on the work of historians who applied the method of discourse analysis to their research. This means that they are not interested in answering questions such as 'Did politization take place?' or 'Were politics involved in the decision-making process?' with a simple 'Yes' or 'No.' An attempt to provide such a simple 'Yes' or 'No' answer often implies that one is becoming an active part of this political discourse. The approach of discourse analysis, however, allows us to zoom out of this debate. It allows us to put some analytical distance between the researcher and the object of research—and from this bird's-eye view, we might be able to see dynamics and interconnections which tend to remain invisible to those actually engaged in the debates.

Such a meta-level perspective therefore changes the kind of questions we ask. Instead of enquiring: 'Did politization take place with regard to a specific project?', a researcher engaged in discourse analysis would ask broader questions, such as: 'Why do the actors in the nuclear waste debate frame politization as something fundamentally powerful and unmanageable?' or 'Why is it considered inherently negative if politics has a central role in decision-making in nuclear waste management?'. These questions, in fact, bring us straight to the broader question that informs this chapter, namely: 'How is the debate about nuclear waste management embedded in the socio-technical and political history of the nuclear enterprise?' The focus on this guiding question implies a certain structure; consequently, this chapter is organised according to this structure. First, we will take a glimpse at how historians have unpacked the multitude and interconnectedness of actors, technologies, politics, events, structures, and dynamics which shaped the 'nuclear enterprise' over the last century. To be sure, this chapter cannot provide a comprehensive history of the nuclear age, which indeed is still waiting to be written. However, by sketching some of the most prominent research trends in the academic field of nuclear history, we will be able to shed light on the complexity of this topic—and we will start to understand why 'politization' and 'politics' are such central topoi of the nuclear discourse. In a second step, we will zoom in on our specific object of interest: nuclear waste. We will do this again through the lens of historical research. This historical perspective allows us to understand that the 'management of nuclear waste' cannot be reduced to a sum of 'technical factors' ranging from the level of ionizing radiation on the surface of a CASTOR® container to the geological characteristics of a certain host rock. Within public, scientific, and administrative nuclear discourse, these 'technical factors' are never simply technical, they are always, in many ways, social as well—and given the history of the nuclear enterprise, always political. This is also why researchers use the term 'socio-technical' to make clear that 'society' and 'technology' are not separated entities but co-created by each other—part II of this book covers this co-creation in depth.

Given the socio-technical co-creation of nuclear waste management regimes, it is anything but surprising that nuclear waste and nuclear waste management projects become politicised, because they always were inherently political. As the following pages will show, the complex and inherently political background of today's nuclear waste debates is brought to the fore by research in the field of nuclear history. Deadlocks in, or the failure of, certain nuclear waste management projects today partly seem to arise from ignoring the history of the nuclear enterprise and its broad social, political, and cultural context. It is therefore of the utmost importance to understand the larger context of the public, scientific, and administrative nuclear discourse and its socio-technical and political history when we aim at safe and acceptable pathways for nuclear waste management.

12.2 Nuclear history: attempts to map highly complex and interconnected fields

Nuclear History is a research field which has grown rapidly in recent years, and it has also become more and more diverse in its research topics (Kalmbach 2017). Today, topics covered by researchers in nuclear history include not only the 'classic' aspects of Cold War deterrence, famous nuclear scientists and their discoveries, and national nuclear energy programs; contemporary historians also prominently address the environmental impact of the nuclear enterprise as well as its impact on (the land rights of) indigenous communities; they research how specific memories of the nuclear enterprise are preserved and conveyed in connection to specific nuclear sites; they investigate the role of nuclear scientists within international organizations; they study how utopias of counterculture were erected on the foundations of antinuclear protest sites and practices; and much more. In doing so, they have increasingly stressed that the civil and the military uses of nuclear technology need to be considered together to fully understand the central role of things nuclear in the 20th century—and the lasting heritage of this nuclear age. Yet, treating the military and civil uses of nuclear technology as two sides of the same coin goes far beyond focussing only on their technical and political interconnections in terms of plutonium-producing reactors and proliferation risks. Beyond these aspects, historical research into the nuclear age has shed light on the interconnectedness of civil and military nuclear applications in terms of the sublime aura in which both applications have been discursively embedded: the common framing is that the utmost destructive and the utmost benign forces are represented by one and the same technology. This image has been created for nuclear bombs (they can destroy the planet—but through deterrence strategies, they also keep the adverse global powers in check), as well as for nuclear reactors (they can produce 'electricity too cheap to meter'—but also cause the worst long-term polluting accidents) and not least for the impact of radionuclides on the human body (their ionizing radiation can kill people in most terrible ways—but it can also save human lives, by means of radiation therapy, from the most dreaded illness of the 20th century, cancer). The discursive dynamics surrounding the civil and military applications of nuclear technology have mutually reinforced each other and co-shaped the emotional backdrop against which meanings are assigned to these technologies.

To somebody unfamiliar with discourse analysis, this all might sound quite abstract, maybe even strange: 'meanings being assigned to a technology'—should this maybe imply post-factualism? Rest assured, it does not; on the contrary. Researchers in the humanities sometimes use complex phrases to underline the complexity of their research topic. And when we use formulations such as 'meanings being assigned to a technology,' we want to underline that technology development as well as knowledge production are highly complex endeavours (Kalmbach 2021). There is no *a priori* knowledge about a technology—no *a priori* answer to the question of whether this technology is 'good' or 'bad'. People develop certain uses of a technological artefact—and in this way they inscribe societal notions into it: norms, ethics, emotions, politics, power dynamics, and the like. At the same time, the use of the technological artefact changes these norms, ethics, emotions, politics, power dynamics, etc. This is why researchers (particularly in Science and Technology Studies[2]—a research field that contemporary nuclear historians draw heavily upon when developing their research questions) use the term 'socio-technical,' because it combines the co-production and mutual shaping of 'society' (including its underlying cultures) and 'technology' (including its underlying sciences) into one single term.

After this short excursus on the term 'socio-technical' (which hopefully also clarified the very title of this article), let us go back to our specific research field of interest: nuclear history. As laid out at the beginning of this section, nuclear history

[2] Science and Technology Studies (STS) is the research field which investigates the social (cultural, political, etc.) foundations of science and technology. This research field is also referred to as 'Social Studies of Science and Technology' or 'Science, Technology, and Society'. It is highly interdisciplinary and transdisciplinary. The websites of the discipline's largest research networks '4S' (https://www.4sonline.org/) and 'EASST' (https://easst.net/) provide first-hand insights into this rapidly growing research community.

is currently researched by academics whose backgrounds are in various subdisciplines of History. It is this interconnection of research fields which are actually often quite separate (sometimes, they may not even be located within the same department) that makes nuclear history such a vibrant research field. Thus, nuclear history operates at the crossroads of the following subdisciplines of academic historical research: International History (deterrence politics, international nuclear organisations, science diplomacy), the History of Science (scientists, scientific networks, processes of knowledge production), Environmental History (pollution caused by bomb tests/nuclear accidents/uranium mining/nuclear waste, water issues, land use), Political History (nuclear proponents and opponents, party politics, energy politics), Cultural History (memory work at nuclear sites, nuclear elites, the counterculture of the antinuclear movement), the History of Technology (reactor design, safety principles, national nuclear programmes), Social History (technocracy, participation, new social movements), Transnational History (Pugwash, NGOs, public understanding of nuclear science), Global History (uranium trade, nuclear refugees, and displaced people), and Economic History (business models, subsidies, and insurance)[3].

One can easily imagine that it is very difficult to combine all these various aspects—which have, of course, taken on very different shapes in different places at different times—into one 'comprehensive history of the nuclear age.' Therefore, as I mentioned above, this comprehensive book is still waiting to be written. However, in recent years, historians have started to make strong efforts to put these various aspects into dialogue. This is done either through developing research questions that link various subdisciplines of academic historical research: my own book on Chernobyl (Kalmbach 2021), for instance, is a combination of Environmental, Political, Social, Cultural, and Transnational History and the History of Technology. Another way of establishing links between these various subdisciplines is by developing concepts that can be applied in as many as possible of these subdisciplines. At the moment, the concept of 'nuclearity'[4] coined by Gabrielle Hecht (2006, 2014) is probably the most popular of these kinds of concept.

As one can tell from the examples of pertinent topics provided in the parentheses to the subdisciplines listed above, 'nuclear' can, in most cases, stand for both civil and military applications. I mention this explicitly to highlight that within academic research, the two 'strands' of nuclear technology are increasingly considered together—because they both inform the nuclear age. The nuclear age thus consists of a highly complex interconnection of all the various topics mentioned above. The fact that 'politics'—in the broadest sense, including everything from party politics to geopolitics, from energy politics to land-use politics—builds the nodes of this web of

---

[3] Because of the large number of publications originating from the field of nuclear history, I am refraining here from referencing individual works. For a recent overview of state-of-the-art research in nuclear history, see Kalmbach (2017).

[4] 'Nuclearity,' in Gabrielle Hecht's use of this term, specifies the discursive counterpart to 'radioactivity.' Whether an object is radioactive or not depends on its materiality. However, if it is 'radioactive,' additional political, economic, social, and cultural processes are required to apply specific meanings to this 'radioactivity' and thus make the 'nuclearity' of the object visible in public discourse.

interconnections makes clear how inherently political everything nuclear actually is. And thus, of course, this also applies to nuclear waste.

12.3 Nuclear history and nuclear waste: the slow emergence of a linchpin topic

Given the centrality of nuclear waste within nuclear debates, it is surprising that historical research on nuclear waste is rather scattered. To underpin this statement, let us consider the example of Germany. There are numerous books that investigate German nuclear history, from the seminal work of Joachim Radkau (1983) to the recent publication by Dolores Augustine (2018), and the dissertations by Stephen Milder (2017) and Andrew Tompkins (2016)—the latter two focus on transnational perspectives. However, when it comes to the specific topic of nuclear waste, there are basically two historians who have focused their research on this topic: Anselm Tiggemann (2004) and Detlev Möller (2009). Astrid Kirchhof and Jan-Henrik Meyer have also recently started to focus on nuclear waste, and first results of their work have been published (Kirchhof 2022, Meyer 2021).

Of course, amongst researchers from the humanities and social sciences, it is not only historians who have been interested in the German nuclear enterprise. With the rise of the antinuclear movement in the 1970s and 1980s, political scientists and sociologists, too, found in these contestations a fertile terrain for research. The books by Dieter Rucht (1980) and Dorothy Nelkin and Michael Pollak (1981) became particularly influential. But as with nuclear history, it was rare for political scientists and sociologists who researched the German case to focus on nuclear waste alone[5]. This has recently changed. A book edited by the ENTRIA[6] research group of the Freie Universität Berlin which locates the German case in an international comparison with various other countries (Brunnengräber *et al* 2015) is a telling example in this regard.

I mention these publications by sociologists and political scientists to show that nuclear history does, of course, not operate in a vacuum. But it is also still true that when we zoom out of the German case and take a look at what has been written about the history of nuclear waste in other countries, the resulting list of books and articles is quite short. The most important work is probably Jacob Darwin Hamblin's history of nuclear waste disposal in the oceans (Hamblin 2008). Furthermore, there have been attempts to historically unpack national waste

[5] This does not, however, imply that from an international perspective, social science research into nuclear waste has been rare; on the contrary (see Salomon *et al* 2010). In particular, researchers from the field of Science and Technology Studies have investigated this topic intensely (see, for instance, Sundqvist 2002). And when it comes to national cases, the US case (Macfarlane and Ewing 2006) and the French case (Barthe 2006) have been researched in depth.

[6] The acronym ENTRIA stands in German for 'Disposal Options for Radioactive Residues: Interdisciplinary Analyses and Development of Evaluation Principles.' ENTRIA was a five-year project (2013–2017) carried out by 12 departments and institutes from German universities and major research institutions and one partner from Switzerland. The disciplines represented within the ENTRIA team, which consisted of more than 60 researchers, were natural sciences, civil engineering, philosophy, law, political sciences, and technology assessment.

governance regimes as well as to compare them (see, for instance, Högselius 2009, 2010, Kasperski and Storm 2020, Lehtonen *et al* 2022). However, as I argued in my article on the state of the art of research into nuclear history (Kalmbach 2017, p 66): 'In comparison to the work undertaken in the social sciences and in particular by STS scholars, historians' contribution to the research literature on nuclear waste is marginal. Even on issues that lie at the heart of their interest, like the construction of memory—which the OECD's Nuclear Energy Agency (NEA) has made a key topic in their strategies governing high-level waste disposal sites (OECD-NEA 2015)—historians barely play a role.' This research lacuna is a problem, because it is indeed historians who are in the best position to 'investigate the role nuclear waste has played in pro-nuclear arguments and the techno-political regimes that allowed the stockpile of high-level waste to grow further and further, despite the fact that no single long-term repository for high-level nuclear waste exists anywhere in the world. They could also question the very definition of what has been considered nuclear waste and what not, in specific geographic areas or scientific networks at certain times, and look for similarities and differences, changes and continuities over space and time. So far, the very nuclearity of nuclear waste lacks historization' (Kalmbach 2017, p 66).

However, even without the many aspects that nuclear historians still need to research regarding nuclear waste, I hope this section has made one thing clear: producing, classifying, managing, handling, researching, administrating, and remembering nuclear waste is happening in the context of the broader history of the nuclear age—a history that is inherently political.

12.4 Conclusions

This chapter showed that it is barely surprising that politics and politization have played and continue to play a major role in nuclear waste management because the historical context in which nuclear waste management is embedded is inherently political. And it also showed that calls for an 'ultimate scientific truth' will barely help us to find solutions to current deadlocks. 'Doing science' is a social process, and thus, of course, politics plays a role. Trying to create the image of a science which is hovering above society and delivering the ultimate right answers which then 'simply' have to be implemented by political decision-makers will most likely not be the way forward for the implementation of safe, sustainable, and (environmentally) just solutions for the long-term disposal or management of high-level nuclear waste—because it is obvious that politics does play a role in these decision-making processes, and actually needs to play a role. And as history prominently reminds us: this politics, of course, changes over time. Decision-making processes that were considered politically legitimate in the 1960s might not pass this bar again today. Our societal understandings of 'just' and 'acceptable' have changed, as did our concepts of 'participation.' And they will continue to change in the future. Given the long timescales over which nuclear waste management regimes will operate, in order to be 'future proof,' these regimes have to make sure that they are able to incorporate and seek to anticipate at least some of these changes—and thus they have to be political. The way forward in nuclear waste management could thus actually lie in a collective effort by all actors involved to critically reflect on the 'politization' and the role of 'politics' in past and present nuclear waste management regimes. And instead of trying to artificially ban politics from the playing field, nuclear waste management actors themselves might actually want to become political agents of change so that they can make their knowledge about what it means to deal with complex socio-technical problems over extremely long timescales available to other fields—because the Anthropocene requires us to learn from these experiences (Ialenti 2020). Therefore, reflecting on the socio-technical and political history of the nuclear age not only allows us to collectively better understand the way 'politics' is tightly interwoven into things nuclear, but also individually, for each actor involved in the nuclear waste discourse, this historical reflection can serve as a prerequisite for finding one's own place in this complex web of socio-technical interconnections—and responsibly filling this place.

Bibliography

Augustine D 2018 *Taking on Technocracy: Nuclear Power in Germany, 1945 to the Present* (New York: Berghahn)

Barthe Y 2006 *Le pouvoir d'indécision. La mise en politique des déchets nucléaires* (Paris: Economica)

Brunnengräber A *et al* 2015 *Nuclear Waste Governance. An International Comparison* (Wiesbaden: Springer VS)

Hamblin J D 2008 *Poison in the Well. Radioactive Waste in the Oceans at the Dawn of the Nuclear Age* (New Brunswick, NJ: Rutgers University Press)

Hecht G 2014 *Being Nuclear. Africans and the Global Uranium Trade* (Cambridge, MA: MIT Press)

Hecht G 2006 Nuclear ontologies *Constellations* **13** 320–31

HoNESt (History of Nuclear Energy and Society) 2021 Full publication list of this international research project at www.honest2020.eu/publications

Högselius P 2010 The decay of communism: managing spent nuclear fuel in the Soviet Union, 1937–1991 *Risk, Hazards and Crisis in Public Policy* **1** 83–109

Högselius P 2009 Spent nuclear fuel policies in historical perspective: an international comparison *Energy Policy* **37** 254–63

Ialenti V 2020 *Deep Time Reckoning. How Future Thinking Can Help Earth Now* (Cambridge, MA: MIT Press)

Kalmbach K 2021 *The Meanings of a Disaster. Chernobyl and Its Afterlives in Britain and France* (New York: Berghahn)

Kalmbach K 2017 Revisiting the nuclear age. State of the art research in nuclear history *Neue Politische Literatur* **62** 49–69

Kasperski T and Strom A 2020 Eternal care. Nuclear waste as toxic legacy and future fantasy *Geschichte und Gesellschaft* **46** 682–705

Kirchhof A 2022 Embodying fear and toxicity: environmental protests against West Germany's final repository for nuclear waste in Gorleben, 1977–1980 *Hazardous Time-Scapes. How to Study Toxicity and Pollution from Multiple Timed, Spaced, and Embodied Perspectives, Ecology and History Series* ed S M Müller and M B Ohman Nielsen (Athens, OH: Ohio University Press) in preparation

Kirchhof A 2018 East–West German transborder entanglements through the nuclear waste sites in Gorleben and Morsleben *J. Hist. Environ. Soc.* **3** 145–73

Kirchhof A and Meyer J-H (ed) Global protest against nuclear power. Transfer and transnational exchange in the 1970s and 1980s *Histor. Soc. Res.* **39** 165–90

Lehtonen M, Cotton M and Kasperski T 2022 Trust and mistrust in radioactive waste management: historical experience from high- and low-trust contexts *Engaging the Atom. The History of Nuclear Energy and Society in Europe from the 1950s to the Present* ed A Kaijser, M Lehtonen, J-H Meyer and M Rubio-Varas (Morgantown: West Virginia University Press) in preparation

Macfarlane A and Ewing R (ed) 2006 *Uncertainty Underground. Yucca Mountain and the Nation's High-Level Nuclear Waste* (Cambridge, MA: MIT Press)

Meyer J-H 2021 Kleine Geschichte der Atomkraft-Kontroverse in Deutschland *Aus Politik und Zeitgeschichte* **71** 10–6 https://www.bpb.de/shop/zeitschriften/apuz/333362/kleine-geschichte-der-atomkraft-kontroverse-in-deutschland/

Milder S 2017 *Greening Democracy: The Anti-Nuclear Movement and Political Environmentalism in West Germany and Beyond, 1968–1983* (Cambridge: Cambridge University Press)

Möller D 2009 *Endlagerung radioaktiver Abfälle in der Bundesrepublik Deutschland. Administrativ-politische Entscheidungsprozesse zwischen Wirtschaftlichkeit und Sicherheit, zwischen nationaler und internationaler Lösung* (Frankfurt am Main: Peter Lang)

Nelkin D and Pollak M 1981 *The Atom Besieged: Extra-Parliamentary Dissent in France and Germany* (Cambridge, MA: MIT Press)

OECD-NEA 2015 Radioactive waste management and constructing memory for future generations *Proceedings of the International Conference and Debate, 15–17 September 2014, Verdun, France* https://oecd-nea.org/jcms/pl_14962

Radkau J 1983 *Aufstieg und Krise der deutschen Atomwirtschaft* (Reinbek: Rowohlt)

Rucht D 1980 *Von Wyhl nach Gorleben. Bürger gegen Atomprogramm und nukleare Entsorgung* (Munich: C H Beck)

Salomon B D *et al* 2010 Three decades of social science research on high-level nuclear waste: Achievements and future challenges *Risk, Hazard & Crisis in Public Policy* **4** 13–47

Sundqvist G 2002 *The Bedrock of Opinion. Science, Technology and Society in the Siting of High-Level Nuclear Waste* (Dordrecht: Springer)

Tiggemann A 2004 *Die 'Achillesferse' der Kernenergie in Deutschland. Zur Kernenergiekontroverse und Geschichte der Entsorgung von den Anfängen bis Gorleben. 1955 bis 1985* (Lauf an der Pegnitz: Europaforum-Verlag)

Tompkins A 2016 *Better Active Than Radioactive!: Anti-Nuclear Protest in 1970s France and West Germany* (Oxford: Oxford University Press)

Chapter 13

Role of stakeholder involvement in the implementation of radioactive waste management projects

David Brazier, Jo-Ann Facella, Pascale Jana Künzi and Erik Setzman

Many national radioactive waste management projects have suffered failures or setbacks in recent decades because socioeconomic considerations and stakeholder concerns were not sufficiently considered. The 'decide–announce–defend approach,' which focused on experts making decisions based on analyses of technical and scientific facts and then announcing those decisions and technical rationales to stakeholders, was common (NEA 2010). Thus, stakeholders were engaged late in the process and for project decisions they were not prepared to simply delegate to experts. As a result, stakeholder acceptance for such projects was not forthcoming. The responsible authorities and waste management organisations did not anticipate that beyond ensuring safety and technical excellence in project design and implementation, broad public awareness, understanding, and acceptance would also be needed to advance projects, specifically those at the local and regional levels.

Experience and research suggests that a broad range of social and political factors can contribute or detract from acceptance, such as:

- trust and confidence in responsible authorities, organisations, and scientists,
- risk perception related to a hazard of such long duration (see part II, chapter 11),
- considerations of inter- and intragenerational justice, and
- perceived stigma and fears associated with nuclear projects in general and waste management in particular.

Additionally, the larger political context, such as the discussion about energy, nuclear weapons, and proliferation as well as climate policy may also play a part (see part I, chapter 4 and part II, chapter 12). In some countries, the opponents of nuclear energy may not be willing to participate in a site selection process for a deep

doi:10.1088/978-0-7503-3095-4ch13

geological repository as long as the power plants are still operating and creating waste. It has become evident that radioactive waste management facilities can only begin to be built and earn acceptance and legitimacy as a result of dialogue with stakeholders and consideration of their questions, needs, and concerns early on, from design to site selection and the implementation and operation of facilities. While safety remains a priority, radioactive waste management projects must seek to build relationships in which host communities are not merely living with the nuclear waste management facility but ideally welcome it as part of their community.

Much research has been conducted on this subject over recent decades at both national and international levels. In this work, the OECD Nuclear Energy Agency's Forum on Stakeholder Confidence (FSC) has been a long-standing and preeminent forum for reflection, collective thinking, and documentation of the factors that contribute to, or detract from, stakeholder confidence and acceptance. It has led thinking about how nuclear waste management projects can be approached, designed, and implemented in a manner that is more responsive to stakeholders and the public. As some national nuclear waste management projects are now advancing beyond site selection, the work of the FSC is progressively more focused on project implementation and operation and the role of stakeholder involvement in enabling these projects to proceed. This includes the involvement of stakeholders in decision-making on such topics as project and environmental monitoring and reporting, knowledge transfer and management, the involvement of youth, and continued learning and communication to build and sustain broad awareness and understanding in a fast-paced and changing media environment.

The discussion that follows outlines key learning from the work of the Forum on Stakeholder Confidence and some of its members. The authors hope that this sparks further dialogue, questions, and discussion.

13.1 Evolution of understanding

In the late 1990s, a number of national programs for long-term nuclear waste management encountered difficulties and had to be redirected or redesigned and restarted because they lacked broad public support. These setbacks mainly occurred because the evolving societal and political dimensions of these projects were not well understood at the time, or they were even underestimated.

Having started from a technical standpoint, in which project implementers 'design' and stakeholders are 'informed and educated,' it became increasingly apparent that beyond ensuring technical excellence and a strong safety case, more meaningful involvement of stakeholders in the project planning and implementation process is needed for successful implementation. This led to the development of new participative approaches for involving stakeholders.

This set the stage for the creation of a range of initiatives and projects to explore stakeholder involvement in radioactive waste management programs going forward, some of which are listed in box 13.1. The FSC, created by the Nuclear Energy Agency in the year 2000, commenced its inaugural workshop meeting in Paris over three days in August (NEA 2000). This was an important and early forum for the

Box 13.1. Range of initiatives, research, and development projects exploring the role of stakeholders and social and socio-technical considerations

In addition to the FSC, a range of national and international initiatives have helped to advance the exploration and discussion of social considerations related to radioactive waste management and the role of stakeholder involvement. The initiatives outlined below provide an example of the breadth of work on this topic.

- The RISCOM (RIsk COMmunication) initiative originated in Sweden in the late 1990s. Its stated objective was to extend the boundaries of the siting and licensing dialogue beyond the confines of 'experts' working for the operator and regulator and introduced the concept of 'stretching' to meet the needs and concerns of local communities. Over the course of its work, a model of transparency was developed, consisting of the triangle of Truth, Legitimacy, and Authenticity and its relationship with trust building was explored. RISCOM II ran from 2000–2003 and tested the theoretical model in five countries.
- The VALDOR symposium (VALues in Decisions On Risk) was hosted in Stockholm and ran from 1999–2009. The initiative brought together industry professionals and academics to explore risk decisions and the role of values in those decisions in nuclear waste management as well as other socially controversial areas such as genetically modified organisms.
- In 2003, the US National Academies published a 216 page book titled, One Step at a Time: The Staged Development of Geologic Repositories for High-Level Radioactive Waste, which focused on the 'adaptive staging' of repository implementation (National Research Council 2003)
- The European Community has funded a range of large social science research programs in the field of radioactive waste management. Framework Programmes #4–6 included TRUSTNET, which focussed on the social management of risk for various hazardous activities and addressed the credibility, effectiveness, and legitimacy of the regulatory framework. The COWAM (COmmunity WAste Management) initiative focussed on the development of practicable recommendations for improving decision-making processes for the siting of nuclear waste management/disposal facilities, and involved elected local and regional councils, local NGOs, local trade unions, regulators, operators, and experts in these discussions. The ARGONA (Arenas for Risk Governance) Project and OBRA (European Observatory for Long-term Governance on Radioactive Waste Management) are also major program initiatives.
- Moreover, the EC-funded project (International) Socio-Technical Challenges for implementing geological disposal (InSOTEC) 'underlined the importance of widening research on RWM to integrate the social aspects of natural science and technology as well as to identify socio-political requirements. It called for more integrated and interdisciplinary research and development projects on RWM with focus on geological disposal, bringing together researchers from natural sciences, social sciences, humanities and engineering.' (https://cordis.europa.eu/project/id/269906) Consequently, research and development projects funded by the European Commission have increasingly considered socio-technical issues, i.e. issues at the interface of society, natural science, and technology. This is visible in the MoDeRn and Modern2020 projects (https://cordis.europa.eu/project/id/232598, http://www.modern2020.eu/) which dealt with the issue of monitoring. According to the Modern2020 website, monitoring consists of the

'Continuous or periodic observations and measurements of engineering, environmental, radiological or other parameters and indicators/characteristics, to help evaluate the behaviour of components of the repository system, or the impacts of the repository and its operation on the environment—and thus to support decision-making during the disposal process and to enhance confidence in the disposal process.' The projects addressed the issue of accounting for stakeholder requirements as well as for the scientific, technical, and planning needs.

- Examples of national interdisciplinary research projects involving the natural sciences, engineering, humanities, law, social and political science, as well as technology assessment include the German ENTRIA and TRANSENS projects. ENTRIA (www.entria.de) addressed interdisciplinary evaluation principles for different waste management options. In the course of the project, several methods and forms of interdisciplinary collaboration, including collaboration between disciplines that are distant from each other (e.g. between natural sciences on the one hand and social and political sciences on the other) were successfully tested and applied. The ongoing transdisciplinary TRANSENS project (www.transens.de) goes one step further by involving partners from outside the classical scientific community in the process of research into fairness, safety, and procedural issues.
- At the OECD/NEA, the FSC collaborates with the Integration Group for the Safety Case (IGSC), which dedicates its work to scientific and technical questions of geologic disposal. The current focuses of the collaboration are the views and approaches of different actors concerning the management of uncertainties in the siting and implementation of repositories (NEA 2021b).
- The CARL (Citizen-stakeholders, Agencies responsible for radioactive waste management, social science Research organizations, and Licensing and regulatory authorities) initiative, a university-led research consortium, ran from 2006–2007 and culminated in a final report in 2007, titled 'Wanting the Unwanted: Effects of Public and Stakeholder Involvement in the Long-Term Management of Radioactive Waste and the Siting of Repository Facilities.'

exploration of these societal issues and continues to this day to play a leading role in advancing discussions and processes and sharing best practice. The FSC is composed of nominees from NEA member countries, including the representatives of implementers, regulators, policy makers, and research and development personnel with responsibility for, and experience of, stakeholder interaction. The FSC's mandate is to explore and advance the understanding of the emerging societal requirements to advance nuclear waste management programs.

The evolution of the FSC activities mirrors a broader public dialogue and a cultural shift in the understanding of what is required to implement long-term nuclear waste management projects in modern times. The FSC has provided a forum for sharing experience, discussing challenges, and developing best practice in stakeholder involvement. It has contributed to a deeper exploration and evolution of stakeholder approaches, as well as building understanding of societal issues and requirements and encouraging new decision-making processes which fit within the evolving sociopolitical context.

As outlined in the foreword, by the late 1990s there was a growing awareness in the nuclear waste management community that 'technical expertise and technical confidence are insufficient, on their own, to justify waste management solutions to a wider audience, or to see them through to successful implementation' (NEA 2004, p 3). Greater stakeholder confidence and involvement are needed to advance waste management projects: 'Because of changes in society's decision-making environment and heightened public sensitivity to all matters connected with environmental protection, nuclear power, radioactivity, and especially radioactive waste, any decision regarding whether, when, and how to implement waste management solutions will typically require thorough public examination and the involvement of many relevant stakeholders.' (NEA 2004, p 3). This marked a significant change in understanding and approach.

The same report noted that with this awareness came a cultural change—'stakeholder dialogue had become a lead principle in radioactive waste management' (NEA 2004). As outlined in the report, 'The new dynamic of dialogue and decision-making is characterised as a shift from the traditional "decide, announce and defend" model, for which the focus was almost exclusively on technical content, to one of "engage, interact and co-operate," for which both technical content and quality of process are of comparable importance to a constructive outcome' (NEA 2003, p 55).

The understanding that stakeholder involvement and trust are critical to advancing national waste-management programs has led to an exploration of how public perception and stakeholder confidence are formed with respect to radioactive waste management programs, and in particular, their safety. This understanding has also been the genesis of work that has developed new approaches to stakeholder engagement, cooperation, and collaboration and has led to new questions, such as 'How can waste management programs become more responsive to local needs and concerns of people?' Broadening the focus beyond technical issues, the importance of addressing social, economic, environmental, and ethical issues has become a recognised necessity. Intra- and intergenerational equity, justice, and legitimacy have become important considerations, and it is now understood that institutions and decision-making processes need to be structured in a way that ensures they are transparent, trustworthy, credible, and fair.

Some examples of past FSC topics and themes are outlined in box 13.2. These underline the need for stakeholder involvement to help build the long-term relationships and support needed to implement projects. Although the shape and form of stakeholder involvement will necessarily vary by country and community, reflecting different legal and political systems, cultures, and histories, these projects can only be implemented with the involvement, and to some extent the support, of the affected communities (and the public), working together with implementers towards a common set of objectives or vision.

Building on this work, the FSC continues to explore and share experience with key questions, such as:

Box 13.2. Examples of FSC topics and themes

- **Facility design:** Understanding that 'the people living in the host community must be willing to live with and help maintain the facility over the years' (NEA 2008c), the question of how a facility can be better integrated with the host community and be made attractive for the long term becomes critical. Building a sustainable relationship between the facility and the community becomes an important objective and the understanding of how to do so becomes a key question. In 2008, the FSC began to explore how waste management facilities could become a durable and attractive part of the fabric of the local community (NEA 2008c).
- **Decision-making processes:** Decision-making also came under review. No longer viewed as a relatively short-lived and straightforward activity which can be completed within a single generation, the implementation of long-term radioactive waste management projects came to be viewed as requiring an extended and incremental process, taking several decades to complete, over which, the needed relationships could be built (NEA 2008b). In this context, decision-making processes need to be designed to be inclusive and to embody the societal values that people hold to be important, accommodating potentially competing values and goals in order to obtain the broad societal approval and support needed by these projects. Individual circumstances and learnings were explored in a series of national workshops (NEA 2008a).
- **Empowerment of stakeholders in decision-making** (NEA 2010): Growing experience suggests that stakeholders need to be empowered. The increasing need to empower stakeholders in radioactive waste management processes over the first decade of the 21st century was documented in a 2010 anniversary publication which characterised the evolution as a leap 'from Tokenism to Real Participation.' The report outlined important changes that took place in citizen participation for radioactive waste management over the previous decade. It highlighted 'a shift from information and consultation towards citizen influence and power, mirroring Arnstein's ladder of participation[1], and a shift from overt conflict or resigned acceptance to volunteering and collaboration by local communities.' The need for community empowerment measures, socioeconomic benefits, and a range of administrative formats for collaboration were recognized and discussed.
- **Words and symbols** (NEA 2011): In exploring how to effectively communicate with stakeholders about the project, the role of symbols in building understanding is an important consideration and became an area of FSC study. The work explored how 'key concepts of RWM [radioactive waste management], such as safety, risk, reversibility or storage and disposal, may carry a variety of meanings or symbolic connotations.' These need to be recognized, understood, and addressed in decision-making. Stakeholder involvement and a dialogue with stakeholders can bring out these often hidden meanings.

[1] Sherry Arnstein (Arnstein 1969) created the 'Ladder of citizen participation' to show different levels of citizen participation, starting on the lowest rung of the ladder with (1) manipulation, moving up to (2) therapy, (3) informing, (4) consultation, (5) placation, (6) partnership, (7) delegated power, and (8) citizen control. With each step up on the ladder, citizen power increases in the decision-making processes (from non-participation to tokenism and citizen power).

- What is the role of local versus regional stakeholders in advancing projects?
- How can transparency in decision-making be strengthened through stakeholder involvement processes?
- What are the different types and levels of stakeholder involvement?
- How can youth be empowered and intergenerational outreach be achieved?
- How can 'added value' be developed for the affected communities?

13.2 Factors that build and maintain confidence

The process of stakeholder involvement as it relates to the implementation of radioactive waste management projects can be compared to a journey with many steps or phases, the ultimate aim of which is to build broad confidence in the radioactive waste management project. In the case of deep geological repositories for nuclear waste, these numerous steps can include initial contact and the building of general awareness about the project. The necessary steps also include feasibility studies, site selection, construction, operation, closure, monitoring, as well as knowledge and awareness preservation. Each of these is an opportunity to engage stakeholders in order to build their understanding of the project and for the implementer to understand stakeholder needs and address them as far as possible through the advancement of the work. The ultimate goal is stakeholder confidence in the safety of the facility and the disposal method in the short and long terms, while supporting, if not enhancing, the social and economic well-being of the host community.

Experience suggests that a broad and inclusive approach should be taken to the definition of 'stakeholder.' A broad range of stakeholders needs to be involved, including the dissenting voices of stakeholders who are neither keen on the journey nor the final destination. The FSC was inspired by the Aarhus Convention's definition of 'the public concerned' (UNECE 1998) and thus views a 'stakeholder' as 'any actor—institution, group, or individual—with an interest or a role to play in the radioactive waste management process' (NEA 2013, p 47). According to the NEA (2013), this includes:

- the general public;
- demographic groups (such as young people);
- residents;
- the representatives or elected officials of local communities;
- national/regional government ministries/departments;
- regulators;
- national/local NGOs,
- local pressure groups (that could be either for or against a given project);
- trade unions;
- the media;
- the scientific research community;
- the implementing organisation;
- the nuclear industry;
- contractors;

- waste producers;
- international organisations;

Not included on the list, but also recognized as important stakeholders are:
- landowners,
- neighbouring countries (governments, ministries, etc.), especially when nuclear facilities are sited near national boundaries.

An example of stakeholder groups involved in, or affected by, nuclear waste management in Sweden is given in figure 13.1. In engaging these stakeholders, 'one size' does not fit all. Each of them may have different needs with respect to their engagement and different expectations about their roles and responsibilities, which may be supported by national and international legal provisions. Even within a single group, needs and expectations may vary throughout the process. Depending on the stage of the journey, different stakeholders will be involved and at different levels (from information and consultation to genuine and meaningful participation).

This journey may take decades from planning to completion and while a roadmap with a timetable is necessary, unexpected detours, setbacks, and uphill tasks have to be expected. Experience suggests that this takes endurance, sufficient resources, and patience on the part of all the participating actors, in particular, because the stakeholders engaged will have differing roles and levels of knowledge, backgrounds, and a diversity of (world) views. Along the way, participants will get to know each other and—ideally—learn collectively. Mutual trust and understanding has a chance to grow with time spent together and has to be guarded, as it can be lost quickly. Because it requires so much time to reach the destination, participants will drop out, and there is a risk that knowledge will be lost. New participants require resources so that they can be integrated, but they also add fresh views and perspectives, which may benefit the overall process.

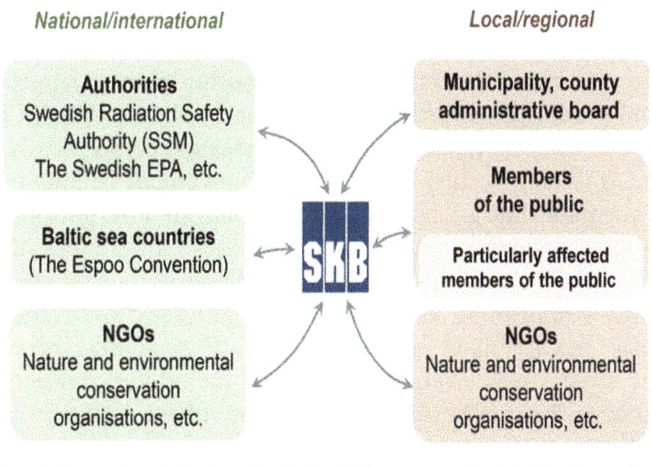

Figure 13.1. Overview of the main stakeholders in the Swedish spent nuclear fuel final repository consultation and the EIA process (reproduced by permission of The Swedish Nuclear Fuel and Waste Management Co (SKB)).

The roles, responsibilities, and tasks of each stakeholder as defined early in this journey may need to evolve over time to sustain their involvement and the participatory process overall. The engagement process must therefore be flexible and adaptive enough to accommodate these changes. For instance, as the process advances and stakeholders mature, some may seek to take on stronger leadership roles within the process. Potential host communities and siting areas, in particular, may come to take a strong role in directing activities (both research and engagement) as a means to address their needs. To the extent that the engagement process seeks to be collaborative and dialogue driven, shifting roles can be expected over the course of the engagement. As the emphasis on collaborative approaches increases, neither the implementer nor an individual stakeholder can control the outcome of engagement activities in which many stakeholders are empowered to shape outcomes. This necessarily introduces uncertainty into project outcomes and time frames, creating opportunities for current participating stakeholders, new entrants, and/or project opponents to either support, delay, or undermine engagement activities. New stakeholders will continue to enter the dialogue and some participants will drop out for various reasons throughout the process and the project implementation (due to frustration, disagreement, exhaustion, age, limited terms in office, etc). A dialogue that addresses issues and concerns, an openness to change, and an ability to adapt may help to retain the engagement of stakeholders, successfully incorporate new entrants into existing engagement activities, minimise the impact of opponents who wish to block or disrupt engagement activities, and strengthen confidence overall.

This metaphor of a 'journey' may help us to grasp the complexity of a participatory process in nuclear waste management, especially in the long journeys that lead to site selection for deep geological repositories and some near-surface repositories. With the recent growth of projects related to the decommissioning of nuclear facilities and legacy management, some different stakeholder needs and concerns may arise.

While some countries are still preparing their journey, other countries are already on track or are nearing the end of their journey towards implementation. These projects differ, and each is unique, depending on the project itself (deep geological repositories, the decommissioning of nuclear power plants, legacy management, etc.), and also on the legal, cultural, historical, and political context in each specific country. In some countries, participatory processes have specifically been designed for this purpose, and as a result, stakeholder involvement is sometimes taken to a new level with formats that go beyond the national or international legal requirements (the Aarhus convention, etc). Despite all these differences, there is a strong agreement within the FSC and the international community that specific factors and principles require attention in order to gain legitimacy or a social license[2] for radioactive waste management projects. These factors and principles create a foundation that strengthens overall confidence, including confidence in the decision-making process, the institutions and people involved, and the transparency and

[2] Having a 'social licence' refers to the situation in which a project has ongoing approval within the local community and among other stakeholders, and also has political and public acceptance (IAEA 2021).

fairness of the process as well as the safe planning, implementation, and operation of a radioactive waste management project.

Transparency: Transparency is a fundamental pillar that touches various aspects, such as a project's goals, process, criteria, structure, communication, and behaviour. This implies being open and honest regarding the communication and dialogue engaged in by the project, including everything from its aim and purpose, financing, roles, time schedules, plans, siting, design, negative or positive impacts and risks to current changes and progress (NEA 2013, pp 57–60, NEA 2014, COWAM 2007).

Behaviour: 'Those involved in waste management need to be: respectful of each other's roles; competent and rigorous in their field of expertise; open, transparent, and willing to involve others; pro-active in initiating dialogue; willing to listen to, and respond to, a broad variety of stakeholders' (NEA 2008a, p 1). Experts interacting with the general public and local stakeholders thus need strong skills in communication, authenticity, and the will to engage (NEA 2021a). Words and deeds need to correspond—in particular, those that originate from the authorities and the implementing organisation. Mutual tolerance is just as important as the ability to accept ambiguities and differences of opinion. An important ability is to recognize that emotions are often tied to arguments and to react accordingly with empathy.

Early engagement: Early engagement ideally starts with the broad engagement of stakeholders in the design of the process (COWAM 2007). This also means 'that stakeholders should be involved upstream, while options are still wide open' (NEA 2015b). This is often a challenge, because stakeholders may not have sufficient interest to engage until a project moves from a generic to a specific stage. Early engagement may also help to raise awareness of the issue.

Definition of roles, responsibilities, and tasks: These need to be clearly defined and understood from the onset of a process (NEA 2015). This includes a communication of the opportunities and limits that can influence the outcome (COWAM 2007).

Communication (NEA 2017b): The communication has to be clear, consistent, and adapted to specific stakeholder groups. Engaging with local stakeholders and lay people will require a 'translation' of very technical concepts and terms without modification of the content. Interactions between experts and non-experts are encouraged. Underground research laboratories, technology demonstration facilities, information centers, and exhibits may also be good ways to communicate and to help 'build confidence with stakeholders for their understanding of the important processes governing repository performance' (NEA 2013).

Commitment: Decades-long projects require long-term commitment and patience from all the stakeholders involved, but specifically from the organization or institution that leads such a process. Sufficient funding and adequately trained staff are part of that commitment. The questions, needs, and concerns of the affected communities and the public have to be addressed. Funding may be provided to a siting region or affected communities so that they can fulfill their role and for competence building.

Openness and Flexibility: These are fundamental for the process, communication and regarding the attitudes of all involved stakeholders. Important pillars of openness and flexibility include:

- willingness to change and adapt, for example, when new issues/knowledge arise
- open attitude about the outcome
- open and easy access to information (including technical publications)
- detailed technical publications, which are structured in a way that enables stakeholders to oversee and comprehend the content
- opportunities for all interested stakeholders to participate in the decision-making process

Time: Sound planning is important, but detours have to be anticipated. Circumstances may require additional time to integrate new knowledge for collective learning, etc. Generally, local stakeholders need time to get to know the other stakeholders, to learn about their roles and tasks, and to build knowledge and capacity. Time is also a crucial factor that is needed to build trust and confidence. On the other hand, the long timescales may be a challenge or a benefit in terms of the commitment of the stakeholders involved.

Stepwise approach (NEA 2008b, COWAM 2007): In decade-long processes, it is important to have milestones (for examples see figure 13.2). For every phase and milestone of the process, the conditions required to enter the next phase or to withdraw must be described and well known. Based on experience, a process based on voluntary participation and the right of potential host communities to withdraw at any point is preferable. This provides clarity and transparency to all stakeholders. Some flexibility and adaption to inevitable changes need to be built in, since such changes should be expected as knowledge builds (see part I, chapter 4).

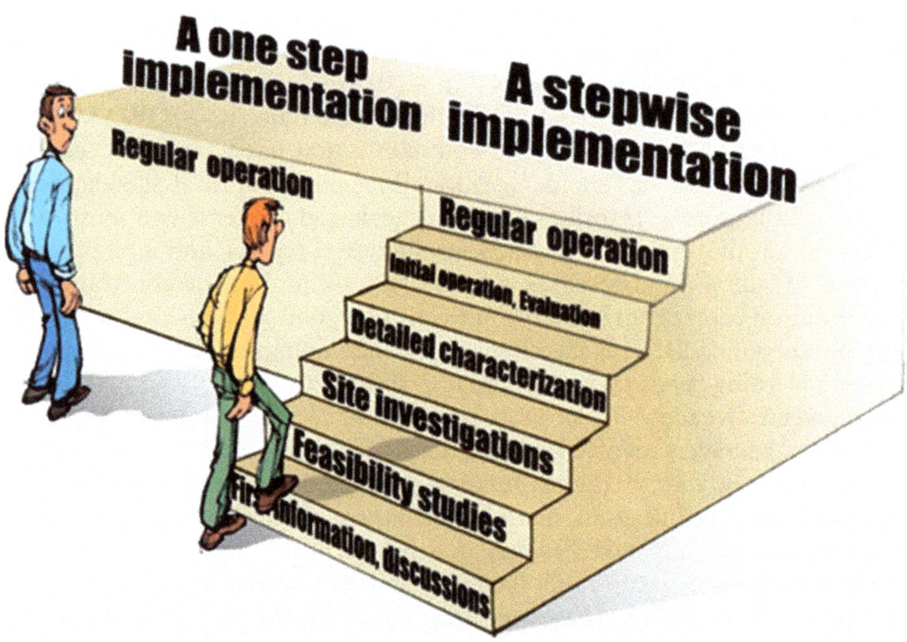

Figure 13.2. The preferred stepwise information, decision-making, and implementation process is of great importance (reproduced by permission of The Swedish Nuclear Fuel and Waste Management Co (SKB)).

Safety: Safety is a preeminent requirement for any radioactive waste management project. Technical questions and issues must be addressed to demonstrate the long-term technical safety of the project to regulatory agencies so that projects can be licensed. Different considerations may exist regarding risk and safety amongst the different stakeholders. Societal concerns regarding safety need to be heard and addressed.

All of these factors can help to create a foundation that strengthens confidence in a radioactive waste management project. This includes confidence in the decision-making process, the institutions and people involved, and the transparency and fairness of the process as well as the safe planning, implementation, and operation of the project.

13.3 Viewing a radioactive waste management project through the stakeholder lens: addressing stakeholder concerns and capacity building

Safety is a preeminent requirement for any radioactive waste management project. Technical questions and issues must be addressed in order to demonstrate the

Box 13.3. Stakeholder concerns

Across multiple national programs, stakeholder questions and concerns tend to touch on a number of themes:

- how to turn a national request into a topic of local interest and avoid 'not in my back yard' (NIMBY) thinking,
- making complicated topics such as radiation, nuclear safety, and high-level scientific language understandable to the lay public,
- the importance of acknowledging values and principles in addition to facts and rational arguments when considering potential impacts on the environment,
- health concerns, worries, and psycho-social aspects related to the project in general,
- skepticism about deep geological disposal as an appropriate waste management option,
- bad will and damage to image, tourism, or the reputation of the community or region,
- falling property and real-estate values which may be caused by the project,
- understanding the extremely long-term time perspective,
- expectations for jobs, the labor market, education, and employment opportunities,
- expectations of positive effects associated with attracting new residents, companies, businesses, and visitors,
- potential technological, economical, and commercial synergies with local companies,
- contribution to a prosperous and flourishing society, local economy, and future,
- potential effects of the transportation of nuclear waste (this may also be a concern to communities along transportation routes),
- preserving records, knowledge, and memory across generations (NEA 2020).

long-term technical safety of the project to regulatory agencies so that projects can be licensed. The questions and concerns of the stakeholders also need to be addressed for projects to advance. Experience has shown that stakeholders have questions and raise issues beyond those of technical experts, and that such questions and issues need to be addressed in order for projects to proceed (see box 13.3).

Many national programs make resources available (information and funding) to support participation in the decision-making processes, in particular, the site selection processes. These programs are designed to ensure that stakeholders and, in particular, affected communities that are involved have the support they need at each step in the process. Capacity building can help to enable a community to learn about the project, reflect on its interest, encourage local discussion and dialogue, and participate in studies throughout the preliminary assessment phase. Programs are often designed to cover the costs incurred by the running of learning groups in the community, the provision of learning and engagement sessions in the community, the cost of staff time, the engagement of third-party experts to address community questions, etc.

13.4 Case studies

The following case studies from Canada, Sweden, and Switzerland are examples of the implementation of stakeholder involvement in those respective countries. The factors and principles mentioned in section 4.2 play an important role in all of these national waste management programs.

Additional case studies from Belgium, Canada, Czech Republic, France, Germany, Hungary, Italy, Japan, Russia, Spain, Sweden, Switzerland, the United Kingdom and the United States can be found in The Forum on Stakeholder Confidence Report on Dialogue in the Long-Term Management of Radioactive Waste (NEA 2021a).

13.4.1 Canada

Prior to the formation of the Nuclear Waste Management Organization (NWMO), much work had been completed toward the development of a deep geological disposal concept for the long-term management of used nuclear fuel. This concept was the subject of a close-to-ten-year environmental assessment review (Seaborn 1998). In 1998, this review concluded that although the concept had been demonstrated to be technically safe for the conceptual stage of development, it had not been demonstrated to be sufficiently socially acceptable to be adopted as Canada's plan.

The 'Nuclear Fuel Waste Act' was brought into force in 2002 to ensure that a safe national solution for the long-term management of nuclear fuel waste was developed, selected by the government, and implemented. The Act required the formation of the NWMO to study and identify a preferred management approach and then implement the approach selected by the government. Building on earlier experience, the NWMO understood that any long-term solution needed to be developed collaboratively with Canadians and Indigenous peoples and initiated a nationwide dialogue to develop a national approach. Adaptive Phased Management (APM, see part I, chapter 7) emerged from this three-year dialogue with specialists and the public. It is based on the values and objectives that they identified. In 2007, the government of Canada selected APM as the country's plan for the long-term management of Canada's used nuclear fuel.

The design of APM reflects its collaborative development, including both a technical method and a management system. The technical method involves the development of a deep geological repository in a suitable rock formation to safely contain and isolate used nuclear fuel. The management system, as outlined by citizens, involves moving towards this ultimate objective in steps through phased and adaptive decision-making, supported by public engagement and continuous learning.

A siting process is underway to select a single preferred site with an informed and willing host. The process began with an open invitation for communities interested in learning about and exploring the project to come forward and become involved in the process, with the option to leave at any time. Initially, 22 communities expressed interest in learning about the project and exploring their interest to host it. Based on a series of progressively more detailed technical and social evaluations, this list was gradually narrowed down to two potential site areas. Up until this point, these areas had been involved in learning about and exploring the project for ten years.

The project will only proceed in an area that can be demonstrated to be safe and has informed and willing hosts. To meet these latter criteria, the siting process requires communities and the NWMO to work together to explore the project and ultimately the ability to move forward together in partnership. This has included a process of ongoing community engagement and activities around key milestones, such as community sustainability visioning, community development of values and principles to guide discussions about the project with the NWMO, and shared project visioning to reflect on how the project might best fit with the community and the area and its potential to help advance towards a future envisioned by those who live there. Most recently, work has focussed on collaboratively developing criteria for assessing community and area willingness and support for the project. This will need to be demonstrated in 2023, after the long period of learning about and exploring the project, in order for the project to proceed in the area.

Work is focused both on municipal and Indigenous communities in the siting area, using community-driven approaches and processes which reflect the sensitivities, culture, and ways of life of each individual community. Throughout this work, the NWMO has committed to contributing to reconciliation with Indigenous peoples.

Local work is led by community liaison committees and working groups which meet regularly and help to advance learning and discussion in the community, and also includes: the establishment of community offices with NWMO staff working and living in the community; the development of standing and mobile information exhibits to foster learning and discussion; the convention of topic-specific workshops and open houses; and regular participation in community and area events and service group and planning meetings.

An important part of engagement activities is exploration of the safety of the project from the perspective of technical experts and also from the perspective of citizens through work to achieve the requirements of 'social safety' as outlined by people in the siting areas. Technical studies are planned and implemented and the findings reviewed in collaboration with people in the area. The identification of specific sites for borehole studies were also selected together with people in the area, factoring in social, economic, and cultural considerations. There is a commitment to interweave Indigenous knowledge with western science throughout this work. Through work with the Indigenous communities involved in the siting process, technical studies are guided by traditional knowledge keepers, local guides, and ceremony (see figure 13.3 left).

A 'Learn More Program' makes resources (information and funding) available to support community participation in the site selection process (see figure 13.3 right). This program has evolved over the course of the site selection process to ensure that communities have the support they need at each step in the process. These resources are intended to help enable a community to learn about the project, reflect on its interest, and participate in assessment studies and also to encourage local discussion and dialogue. Funding is also provided to support activities related to future community decision-making, including access to independent third-party advice to represent community interests in discussions with the NWMO.

Current planning timelines have a single preferred site selected in 2023, at which point, detailed site characterization will begin and the federal regulatory review process will be triggered. Beyond technical safety, social, economic, and cultural considerations are important in selecting a preferred area and site. Considerations include: support for the project among leadership, grassroots, and neighbouring communities; social understanding and confidence in the safety of the project, the safety assessment, and the safety case for the project; and alignment of project

Figure 13.3. Ceremony to implement reconciliation policy (*left*); Ignace community members visit the NWMO mobile exhibit to learn more about the project (*right*) (reproduced by permission of The Canadian Nuclear Waste Management Organization (NWMO)).

well-being opportunities with community vision and strategies. Project design and construction is expected to begin in 2033.

13.4.2 Sweden (based on NEA 2021a, see also part II, chapter 15)

The Swedish Nuclear Waste Management Programme was initiated in the 1970s. In the early 1980s, a joint venture was formed by nuclear reactor owners in order to fulfill the legal requirements on the industry to take care of the radioactive waste disposal, followed by new legislation (for example the Nuclear Activities Act) and the foundation of the implementer Swedish Nuclear Fuel and Waste Management Co (SKB). Early drillings and area investigations undertaken in the 1980s with the purpose of enhancing geological knowledge for the final disposal of spent nuclear fuel met resistance and opposition and raised negative local opinions. This, together with the shortcomings or failures to progress of other national programs, marked the need for new thinking and a new strategy.

At the beginning of the 1990s, the implementer SKB presented its adapted strategy based on principles such as long-term funding, clear roles, responsibilities and process, voluntary participation, stakeholder involvement, dialogue, openness, transparency, and stepwise decision-making.

Close and frequent interaction with potential host communities and local stakeholders has become, and remains, crucial. However, partnerships or direct cooperation between the implementer/industry and potential host communities have been avoided so as not to jeopardize each other's integrity. Municipalities took great part and responsibility from the beginning by informing, inviting, and engaging local stakeholders and setting up the reference and working groups. The implementer SKB was able to start with these groups and enlarge them as necessary. This turned out to be a very good way to avoid missing any important local stakeholders and to ensure the involvement of the most concerned and potentially affected (e.g. nearby residents, landowners, and the municipality). During the site investigations, the Environmental Impact Assessment (EIA) and public consultation procedures provided a valuable core framework for the process of stakeholder involvement, knowledge and capacity building, and stepwise progress. External independent experts were engaged to investigate and explain various environmental and societal impacts, concerns, fears, and effects and to answer questions forwarded by the municipality reference and working groups and from decision-makers. Scientists and researchers could also propose topics and apply for support in the 'Social Science Research Programme,' which ran from 2004 until 2010.

The government participates in the RWM process through the triannual Research, Development & Demonstration (RD&D) program and funding reviews and decisions. The Swedish National Council for Nuclear Waste is an advisory body to the government. It investigates and clarifies matters relating to nuclear waste and serves as a knowledge base for other stakeholders in the area. The Council conducts broad discussions, national and international hearings, and seminars to raise important RWM topics that are not discussed elsewhere.

The Swedish regulator SSM is responsible for the licensing procedures under the Nuclear Activities Act. SSM is also a referral body for the Land and Environment Court and the preparatory authority for the government's decision. SSM reviews the application and is the Swedish expert body in the nuclear safety field, during consultations, EIAs, and licensing under the Environmental Code. SSM participates in seminars organized by the government advisory body and other public meetings within the field. It also has regular meetings with the affected municipalities and NGOs. SSM publishes information, such as news from the review process, on its website. It also documents all meetings with SKB and other operators in a public diary.

In addition to the Government, the regulator, and the Land and Environment Court, the municipalities have quite strong positions in Sweden, both in general, and specifically relating to radioactive waste management due to their Environmental Code veto right and the possible use of referendums, while the regional level has a more administrative profile and is less powerful. Regional authorities have a specific role according to the environmental code, which relates to the consultation process for the use of disposal facilities. In the early stage of the EIA process, regional authorities were also the organisers of regional EIA forums in both the potential host communities, Oskarshamn and Östhammar.

In addition to local citizens, nearby residents, land and property owners, and the public, a wide range of companies and associations continue to be part of the local/municipal reference and working groups. The potential host municipalities and NGOs active in radioactive waste management have received funding to participate in the process from the Nuclear Waste Fund following the EIA and consultation process (allocated by law by the reactor owners). This has included knowledge and capacity building as well as both informal and formal reviews of plans, impacts, and applications and tracking the project and final disposal issues in general. Since 2018, they have received compensation from the State to continue their work. Through the years, SKB has made special efforts to reach out to, for example, women and young people. Such initiatives have offered and continue to offer site and facility visits, open houses, education packages, and school projects as well as approaching relevant associations in their normal meetings and venues.

Since site selection in 2009, there has been agreement about an 'Added Value Programme' (AVP), which was launched by the municipalities involved i.e. Östhammar and Oskarshamn, SKB, and its owners. Value-added initiatives are financed directly by the owners of SKB and are not funded by the Nuclear Waste Fund. The aim is to create mutual benefits for both parties (the municipality and industry) through actions and investments of different kinds treated and decided by a board with members from both parties.

The planned future host municipalities for the spent fuel encapsulation plant (Oskarshamn) and the final repository (Östhammar, see figure 13.4) have now, after a long and fruitful voluntary process (see also part II, chapter 15), both declared and decided that they will accept the facilities. The regulator and the Land and Environment court have given their statements to the Government. They are both positive in principle, but express some reservations about a further need to confirm the integrity of the copper canisters to ensure long-term safety. A Government

Figure 13.4. Illustration of the planned final repository for spent nuclear fuel close to the nuclear power plant in Forsmark Sweden (reproduced by permission of The Swedish Nuclear Fuel and Waste Management Co. (SKB)).

decision on approval of the final disposal system and siting is expected in the near future,[3] meaning that construction could possibly begin around 2025.

13.4.3 Switzerland (see also part II, chapter 15)

To plan and implement final disposal, the National Cooperative for the Disposal of Radioactive Waste (Nagra) was established in 1972. Nagra is responsible for the disposal of all types of radioactive waste in deep geological repositories, including the preparation of proposals and license applications, as well as the operation of the repositories. Site selection began in the 1970s. In 1993, Nagra proposed Wellenberg in Canton Nidwalden as the repository site for low- and intermediate-level waste. In 1994, an application for a general licence was submitted. The review of the licence application by the federal nuclear safety authorities came to a positive conclusion, but the grant of the mining concession required by cantonal legislation was rejected by the voters of Canton Nidwalden in a cantonal referendum (the siting community of Wolfenschiessen voted in favor of the concession). A staged approach was subsequently proposed and an application for an exploratory shaft only (for a modified repository project) was submitted in 2001. The population of Canton Nidwalden rejected the concession application in September 2002, but the commune of Wolfenschiessen voted in favor of the project again. Due to the lack of public support, the project was stopped and the Wellenberg project considered a failure.

As a next step, the Swiss Federal Office of Energy (SFOE 2006) analyzed past experiences in Switzerland and the model of local partnerships in Belgium in order to develop requirements for a future participation process. Additionally, ETH Zürich (2006) published a case study of the Wellenberg project to derive building

[3] After the editorial deadline: On 27 January 2022, the Swedish government approved the final disposal system and siting, cf. https://www.skb.com/news/the-government-approves-skbs-final-repository-system/ [accessed 27 January 2022].

blocks for acceptable decision-making processes. A fair, transparent, participatory, stepwise process with clearly defined criteria, roles, and responsibilities for all stakeholders was then designed. The so-called Sectoral Plan for Deep Geological Repositories—a federal spatial planning instrument—was devised in a consultative way, which included the cantons and neighbouring countries. To get broad feedback, workshops were held for political parties and organisations and focus groups were organized for representatives of the public. Thus, early and broad engagement in the design of the process was enabled. A public hearing took place at the end of this process and the Conceptual Part of the Sectoral Plan for Deep Geological Repositories (Sectoral Plan) was approved by the federal government in April 2008.

The Sectoral Plan specifies a three-stage site selection procedure and it defines the safety-related criteria. The safety of human beings and the environment is of the highest priority. Safety takes precedence over aspects related to land use (spatial planning), the economy, and society. The Sectoral Plan also defines milestones, the roles of all stakeholders, and the mechanisms for collaboration between stakeholders. Each stage follows the same process: a proposal written by Nagra, review by the authorities, followed by a three-month public consultation and, at the end, the decision of the federal government. The Sectoral Plan will end with the approval of the general licence by parliament. This decision is subject to an optional national referendum. Thus, decisions are taken at the national level. The cantons and the communities have no right of veto for a deep geological repository. Instead, the siting regions have the opportunity for collaboration prior to the consultation procedure that takes place at the end of each stage.

The SFOE is in charge of the site selection procedure and for the regional participation, which is an important pillar of the Sectoral Plan in stages two and three. Intensive stakeholder involvement takes place through so-called 'Regional Conferences' (figure 13.5 left), which were established in each of the six siting regions at the end of stage one. At the end of stage two, a reduction from six to three siting regions was confirmed by the federal government. The three remaining regional conferences, Jura Ost, Nördlich Lägern, and Zürich Nordost, are organized as associations and constitute up to 130 members from local authorities, organized interest groups, and the local populations. German delegates are also included, due to the proximity of the siting regions to the German border. Each regional conference has a management board and an administrative office, as well as topical working groups: (1) regional development, (2) surface infrastructure, and (3) safety. Two regional conferences also have working groups known as 'infrastructure communities.' The regional conferences are funded by SFOE. They run their own websites, where they publish all relevant documents and their statements. The members are compensated to attend working group and plenary meetings. Additionally, the SFOE organizes and offers members knowledge, capacity-building sessions, and excursions on various topics, such as radioactivity and radiological protection, visits to the interim storage facility or the rock laboratory at Mont Terri, and topics determined by the members themselves, such as risk communication, spatial planning, or geology (figure 13.5 right).

The aim of regional participation is to take regional interests and concerns into account at all stages. Regional conferences have specific tasks, which are based on

Figure 13.5. Plenary meetings of a regional conference (*left*) and capacity building for regional conference members at the Rock Laboratory, Mont Terri, in Switzerland (*right*) (reproduced by permission of The Swiss Federal Office of Energy (SFOE)).

the milestones of the site selection procedure according to the Sectoral Plan. In stage two, the regional conferences discussed the placement of the surface facilities and published statements containing their questions, recommendations, and demands. At the end of this process, Nagra defined one or two sites for the placement of surface facilities in each siting region, based on the statements of the regional conferences. Additionally, all regional conferences published a statement regarding the implementer's proposal for stage two. With the approval of the Federal Council in November 2018, the areas of the surface facilities were incorporated into the sectoral plan as an interim result. In stage three, the three remaining sites are to be investigated in more detail and compared from a safety perspective by Nagra. The surface facilities will be specified in collaboration with the regional conferences (which include German delegates) and the cantons. Additionally, socioeconomic aspects will be examined in depth and the regional conferences will be asked to formulate measures to foster sustainable regional development.

13.5 Recent developments

In recent years, existing concepts and themes have received increased or renewed attention. The FSC attempted to deepen the understanding of concepts such as 'added value' and to address issues such as youth involvement and intergenerational decision-making and knowledge management and transfer. The COVID-19 pandemic left its mark on stakeholder involvement as well. Virtual participation and 'science in the living room' received a boost. At the same time, questions about how much virtual participation is feasible and possible arose.

13.5.1 Added value

The concept of 'added value' was introduced by the FSC in 2007 in the report (updated in 2015) 'Fostering a Durable Relationship between a Waste Management Facility and its Host Community. Adding Value through Design and Process' (NEA 2015a). Kojo and Richardson (2019) wrote about the 'The Added-Value Approach—clarifying the key concepts' in the European-Community-supported project 'Implementing Public Participation Approaches in Radioactive Waste Disposal' (2011–2012).

Because the term 'added value' (or 'value added') was recognized as an important concept, but was used in various contexts (within NEA and beyond) and without a clear definition, the FSC defined the term and integrated it into its 'Annotated Glossary' and a Flyer 'Added value in nuclear waste management and processes' (NEA 2022a, 2022b):

'The Forum on Stakeholder Confidence (FSC) of the OECD Nuclear Energy Agency (NEA) uses the term 'added value' to describe the contribution radioactive waste management facilities can make to the communities and areas in which they are located beyond the safe and secure management or disposal of radioactive waste. From the perspective of stakeholder confidence, the concept of added value is important as it is associated with helping to build stronger and more durable relationships between the community and the facility or operator. This, in turn, supports long-term relationship building with the implementer and project sustainability. Some waste management programmes identify added value as a right for the community to receive, and a responsibility for the implementer to deliver, given the national importance of the project. Added value is achieved when the facility and/or ancillary program activities are designed, planned, implemented and/or operated in a manner that contributes to the social and economic well-being of the host community and area.' (NEA 2022b).

13.5.2 Youth involvement and sustainable intergenerational decision-making

The long timescale involved in radioactive waste management—from waste generation to disposal—means that the decision-making process can span many decades and governments. For many people employed in the nuclear waste management sector, the lifecycle of the waste extends beyond their careers or 'terms in office.' Many government and industry sponsored initiatives encourage people to study science, technology, engineering, and mathematics (STEM) subjects and pursue careers in the nuclear and waste management sector, to help ensure continuity of knowledge, skills, and competence. However, the continuity of knowledge and interest in radioactive waste at the community level is more vulnerable to erosion. Community groups associated with siting initiatives for waste disposal have tended to attract older (often already retired) people with spare time on their hands, and this can be offputting for younger people.

Youth involvement was a special theme explored in the FSC's 2016 national workshop held in Bern, Switzerland (NEA 2017a). A group of youth (ages 16–24) from the siting regions in Switzerland participated in the discussions to share their thoughts on getting involved in radioactive waste management debates in their regions. Some of the reasons that younger people did not want to engage were because they could not relate to the problem of radioactive waste in their everyday lives. It appeared too remote and abstract, unlike, for example, climate change or sustainability choices which they could relate to their personal lifestyle choices and actions. One suggestion offered by a young participant was to advertise the link between nuclear waste and nuclear power, because young people were more interested in the nuclear power vs. renewables debate. However, most radioactive waste management agencies have taken care not to combine the debates about new nuclear power with nuclear waste. Specific recommendations for the engagement of youth can be found in NEA (2017a, 40–41).

Many countries have recognized that it takes an extra effort to get young people interested and engaged in radioactive waste management projects and are creating new instruments and formats to secure their engagement. Various programs exist to build youth awareness and capacity for future decision-making[4]. Industry- and community-initiated programs can include:

- resources to support youth participation in decision-making;
- briefings and education sessions with students of all ages;
- funding for student research initiatives;
- scholarships to support science education pathways to related careers;
- targeted youth employment initiatives to encourage participation in the implementation of projects; and,
- involvement in local community meetings and advisory and decision-making forums.

Another consideration for youth engagement is to allow youth to design the engagement structure, scope, and aims, which empowers youth with the authority, agency, and resources to bring in new perspectives that may lie outside the interests of authorities. According to one campaigner on a panel session that informed Canada's radioactive waste policy, youth should not only be engaged in deliberations structured and formatted by authorities, reflecting their aims and interests; rather, as the meeting report on the summit states, 'In terms of engaging more youth in these conversations, there must be an indication that youth can lead, affect and design the plans and processes in a more impactful manner' (NWMO 2021).

13.5.3 Influence of the pandemic on stakeholder involvement

13.5.3.1 Local and international stakeholder engagement events during the COVID-19 pandemic

In some respects, the pandemic has enabled broader stakeholder engagement, despite widespread restrictions on travel and group gatherings. Many stakeholder events have moved online and some have been advertised on the internet, enabling people all over the world to participate by self-subscribing to a virtual event for what would otherwise be an inaccessible 'local meeting' (see part II, chapter 16).

At the international level in particular, one of the well-known obstacles to the involvement of more non-professional stakeholders has been the question of who pays the expenses that cover attendee travel, subsistence, hotel costs, and time. These costs are not typically offered to meeting participants, which has meant the exclusion

[4] Examples of this can be found in *Intergenerational connections in radioactive waste management: Involving children and youth across generations* (NEA 2021c).

of those individuals who cannot afford to cover those costs, or those who lose out on their work time. With the rise of remote video participation, those costs have proved less of a deterrent to participation, as people participate from their own homes with no travel time, and in most cases, attendance fees were waived during the pandemic.

The success of this new mode of involvement was negated by the sheer number of seminars and meetings that emerged, and with no buffer of 'travel time' to shield people from overlapping meetings, work calendars were busier than ever, with many back-to-back meetings arranged.

At the national, regional, and local levels, stakeholder involvement continued virtually as well, but with an awareness of the limits of virtual engagement. When stakeholders have a specific task to fulfill, for example, in an ongoing site selection procedure, virtual meetings are not a sufficient replacement. Video conferences/meetings can work when the collaboration format is already established and stakeholders know each other and the area or specific site. In that case, they can even be more efficient, as there tends to be more focus. Nevertheless, it becomes more challenging when in-depth discussions and site visits are necessary.

From the perspective of engaging local communities, experience suggests that virtual meetings and dialogue may have strengths and weaknesses as engagement approaches. Although some local stakeholders may more readily be able to participate in meetings that are virtual and do not involve leaving home, some stakeholders may not have easy access to computers and/or access to the bandwidth needed; alternatively, they may be uncomfortable with participating in this kind of online event. In addition, virtual events lack the kind of community comradeship provided by personal events. The pandemic has also introduced additional demands and strains into communities beyond these projects, which may further compete for stakeholder attention, such as implementing new health and safety protocols, maintaining services in challenging circumstances, etc. Both the constraints on meeting in person and the additional demands put on communities by the pandemic have made local stakeholder involvement more challenging. Additional time may be needed to advance programs under these conditions.

13.5.3.2 Science in the living room

The pandemic has brought science and risk communication into everyone's living room. Almost every night, debates on facts, data, models, and evidence have been headline news. People have become familiar with themes and topics that were previously popular in specialist radioactive waste management circles, such as charts showing projections and predictions of cases, the variation of risk according to different models, scenarios and assumptions, 'worst-case scenarios,' and stepwise decision-making under conditions of uncertainty. People learnt about the interplay between science and decision-making and the role and importance of the social and behavioral sciences; they also witnessed experts, governments, and even countries disagreeing. They also learnt how models can be wrong (FullFact 2021), how public health messages are constructed, and how even simple protective measures can become embroiled in complex debates. An example is the efficacy of mask wearing, which was at first considered irrelevant, but then made compulsory. People observed many errors in both science and policy-making. 2020 has given us both the best and the worst of science. There was 'retraction of false claims, doubling-down on obvious errors, and undignified social media spats' (BBC 2021). At the time of writing, it is difficult to forecast how trust in science, the science–policy interface, or government-sponsored disposal programs have been affected by all the critical analysis of science in the news; however, indications from survey polls show that trust in scientists remains high, while trust in politicians (already low) dropped lower still.

13.5.4 Knowledge management and transfer

Since radioactive waste management projects usually take several decades from development to the construction and operation of a facility, it is crucial to manage and transfer (relevant) knowledge and information pertaining to technical and societal aspects.

Knowledge transfer is also relevant for future generations. They may judge past decisions differently and even want to adapt them, for example, in the case of the decision for the (final) closure of a deep geological repository or if waste is to be retrieved, because there may be options to reuse some of the waste or to manage it differently in the future. Thus, future generations need to be able to gain an understanding about the decision-making process of the past and to understand how and why decisions were taken and based on what knowledge. Local stakeholders and affected communities may play a crucial role in this effort.

Additionally, legal requirements exist in many countries about long-term awareness preservation for deep geological repositories. These require further exploration of how stakeholders—for example, those from the affected communities—can be meaningfully engaged in this effort.

13.6 Outlook

13.6.1 The role of social media

When social media platforms first emerged over a decade ago, organisations involved in radioactive waste management, like other organizations, had little clue about how to use them. With the passage of time and the widespread growth in their popularity, it seems that every organization has a presence on social media, although organizations have different policies and strategies for how to use them. At a minimum, organizations are willing to use social media platforms to 'push' news and information, taking advantage of a 'free' publishing platform which can reach out to thousands of people. Most platforms have free utilities to record useful metrics, such as how many people potentially viewed a post (the number of 'impressions'), how many interacted with it, 'liked' it, replied, or forwarded it.

One of the challenges for waste management organisations and regulators relates to the question of how to respond to false, negative, or critical replies. Platforms such as Twitter have a reputation for being the place to go to when you want a virtual 'fight.' Account holders could spend many hours trying to defend their position or correct high-volume misconceptions or misleading information, and many organizations have concluded that this is not a productive use of their time. However, there are reputational risks for organizations that are seen not to engage in conversations. After all, successful engagement is a conversation, not a monologue.

Some subjects or circumstances are easier or more important to engage with than others. During a nuclear or radiological emergency, for example, social media accounts are essential for authorities to disseminate important information quickly and directly to the public and other agencies (IAEA 2018); in such cases, there is an urgent need for trusted, reliable voices.

The risks of negative or disruptive replies to posts are reduced when the subject is well defined and involves issues such as geoscience, for example, which is a relatively uncontroversial topic compared to a discussion of issues such as radioactive waste policy or repository siting.

Facebook and YouTube have far greater reaches (billions of active users) than Twitter (users numbering in millions). Twitter, although popular with radioactive waste agencies, journalists, opinion leaders, and other 'professional' users, is not growing in user numbers and it is not that popular with the average member of the public.

Social media has also been responsible for a sharp rise in the spread of misinformation. The scale of this problem became apparent during the pandemic, but it had been present for several years in relation to conspiracy theories circulating about anthropogenic climate change. The scale of the challenge is hard to measure for radioactive waste management, but it is not thought to be unmanageable, because unlike, for example, national elections, COVID-19 restrictions, or climate change, the impact on geopolitics or personal lifestyle changes for the general public is not so great with radioactive waste; therefore, it does not tend to attract conspiracy theorists.

However, we live in an era that presents us with, or at least gives access to, an overwhelming amount of information. The WHO has called this the 'infodemic,' in which many people suffer from 'news fatigue'. Trust surveys tell us (Reuters Institute 2019) that two-thirds of people cannot tell the difference between good journalism and falsehoods and are worried about what is real and fake on the internet. Consequently, at the start of 2021, trust in all sources of information (traditional media and social media) was at an all-time low (Edeleman 2021).

It is well recognized that a large proportion of the population has 'poor information literacy' in relation to technical or scientific matters, and more generally, poor media and digital literacy skills. All of these factors should be considered when designing engagement strategies. Organizations can only do so much fact-checking or debunking (and these strategies will only work if the organization is sufficiently resourced and trusted). In the short term, it is possible for agencies or other trusted organizations to provide training for communities on how to improve their critical thinking and critical appraisal skills, how to spot misleading or 'fake news'—in other words, how we can all become better 'fact checkers'. The medium- to longer-term solution must be to try to improve overall general education. Media literacy surveys indicate that Finland is leading the way, starting with primary school education on evaluating 'fake news' (The Guardian 2020).

Over the long term, and perhaps more fundamentally, the use of collaborative processes to create knowledge in which multiple perspectives are considered and a shared understanding is developed with communities may be necessary to advance programs that easily foster fear—such as nuclear waste management programs, in which it is easy to scare people but much harder to unscare them. Such shared understanding can help to entrench learning and facts, and create some resilience to fake news and fearmongering.

13.6.2 Stakeholder involvement after site selection

After site investigations and site selection have taken place, and in particular, when applications for the required permits have been submitted, the licensing process really begins and the stakeholder involvement enters a new and rather different phase. During the licensing process, the roles and responsibilities of the participating stakeholders tend to become more distinguished, formal, and distanced, at least at some levels. The initiative moves partly from the implementer to the regulator, other authorities, the Government, and the host municipality, as they are, or will be, responsible for the licensing procedures and the final approval or decision-making. Direct cooperation or interaction between stakeholders, for instance between the project implementer and affected communities, may become even more sensitive than before, in order to limit any appearance of undue influence on the part of the implementer. The licensing process should be expected to be (at least to some extent) controversial, complicated, and time consuming, since it is intentionally designed to be a challenge process in which both project supporters and critics are encouraged to participate. Continued efforts to find proper ways of securing contacts, dialogue, and local acceptance are then of the utmost importance in order to maintain trust and confidence through this very public assessment and challenge process.

In addition to reviews, hearings, meetings, and seminars within or connected to the progress of the licensing procedures, it is important for the main stakeholders to remain present, available, and supportive, for example, by supplying information and responding to questions. To the extent required by circumstances, active participation in daily local life and debate over time is still recommended. Activities of mutual interest which may help to sustain confidence and trust during this period can include:

- Communication and information about the status and progress of the licensing process and the community/municipality decision-making process.
- Dialogue concerning further facility planning, design, and environmental precautions if ongoing in parallel with licensing.
- The contents, progress, and results of any added value program.
- Open houses/facilities and public seminars on interesting topics and themes concerning the project, community development, and the planned socio-economic coexistence.

The closer to the final decision-making and approval (or disapproval) a radioactive waste management project comes, the more media coverage, political tension, and social media attention should be expected. Stakes will be high, facts may be challenged, values may be in dispute, but nevertheless, decisions are needed to advance these socially necessary projects. Attention may shift more to the regulatory licensing approval process, and regulators must ensure that they maintain their integrity and independence and do not yield to 'regulatory capture.' Stakeholders must be prepared for this phase. All involved stakeholders should keep in mind how important openness, transparency, honesty, integrity, reliability, participation, and dialogue are, even at this point.

Bibliography

Arnstein S R 1969 A ladder of citizen participation *J. Am. Inst. Plan.* **35** 216–24

BBC 2021 *Analysis: Science in the Time of Covid-19* https://bbc.co.uk/programmes/m000tcqy

COWAM (Community Waste Management 3) 2007 *Quality of Decision-Making Processes, Decision-Making Processes in Radioactive Waste Governance—Insights and Recommendations*

Edeleman 2021 *Trust Barometer* https://edelman.com/trust/2021-trust-barometer

ETH Zürich (Eidgenössische Technische Hochschule Zürich) 2006 *Entscheidungsprozesse Wellenberg. Lagerung radioaktiver Abfälle in der Schweiz* (Zürich, Switzerland) https://ethz.ch/content/dam/ethz/special-interest/usys/tdlab/docs/csproducts/Band_2006.pdf

Full Fact 2021 *How Media Reporting of Modelling has Shaped Our Understanding of the Pandemic* https://fullfact.org/health/scientific-modelling-covid/

IAEA (International Atomic Energy Agency) 2021 *Nuclear Communicators Toolbox; Social Licence* (Vienna: IAEA) https://iaea.org/resources/nuclear-communicators-toolbox/basics/trust/social-licence

IAEA (International Atomic Energy Agency) 2018 *Report on Int. Symp. on Communicating Nuclear and Radiological Emergencies to the Public*, CN265 (Vienna: IAEA)

Kojo M and Richardson P 2019 The use of the added value approach in siting radioactive waste facilities *Conflicts, Participation and Acceptability in Nuclear Waste Governance. Energiepolitik und Klimaschutz. Energy Policy and Climate Protection* ed A Brunnengräber and M Di Nucci (Wiesbaden: Springer VS) pp 207–28

National Research Council 2003 *One Step at a Time: The Staged Development of Geologic Repositories for High-Level Radioactive Waste* (Washington DC: The National Academies Press)

NEA 2000 *Stakeholder Confidence and Radioactive Waste Disposal* Workshop Proceedings (Paris: NEA) https://oecd-nea.org/jcms/pl_13440/stakeholder-confidence-and-radioactive-waste-disposal

NEA 2003 *Learning and Adapting to Societal Requirements for Radioactive Waste Management* https://oecd-nea.org/jcms/pl_13878

NEA 2004 *Learning and Adapting to Societal Requirement for Radioactive Waste Management. Key Findings and Experiences of the Forum on Stakeholder Confidence* (Paris: NEA) https://oecd-nea.org/jcms/pl_13878

NEA 2008a *Decision Making for Radioactive Waste Management. Principles, Action Goals, Confidence Factors* (Paris: NEA) https://oecd-nea.org/jcms/pl_27158/decision-making-for-radioactive-waste-management-english-a4–2008

NEA 2008b *Stepwise Approach to the Long-Term Management of Radioactive Waste* (Paris: NEA) https://oecd-nea.org/jcms/pl_27166/stepwise-approach-to-the-long-term-management-of-radioactive-waste-english-a4–2008

NEA 2008c *Towards Waste Management Facilities that Become a Durable and Attractive Part of the Fabric of Local Community—Relevant Design Features* (Paris: NEA) https://oecd-nea.org/

jcms/pl_27175/towards-waste-management-facilities-that-become-a-durable-and-attractive-part-of-the-fabric-of-local-community-relevant-design-features-us-letter-2008

NEA 2010 *Partnering for Long-Term Management of Radioactive Waste. Evolution and Current Practice in Thirteen Countries* (Paris: NEA) https://oecd-nea.org/jcms/pl_14486/partnering-for-long-term-management-of-radioactive-waste

NEA 2011 *More Than Just Concrete Realities: The Symbolic Dimension of Radioactive Waste and its Management* (Paris: NEA) https://oecd-nea.org/jcms/pl_27144/more-than-just-concrete-realities-english-a4–2011

NEA 2013 *Underground Research Laboratories* (Paris: NEA) https://oecd-nea.org/rwm/reports/2013/78122-rwm-url-brochure.pdf

NEA 2014 *Stakeholder Confidence and Transparency in Radioactive Waste Management* (Paris: NEA) https://oecd-nea.org/jcms/pl_27139/stakeholder-confidence-and-transparency-english-a4–2014

NEA 2015a *Fostering a Durable Relationship Between a Waste Management Facility and its Host Community. Adding Value through Design and Process* (Paris: NEA) https://oecd-nea.org/jcms/pl_14966/fostering-a-durable-relationship-between-a-waste-management-facility-and-its-host-community

NEA 2015b *Stakeholder Involvement in Decision Making: A Short Guide to Issues, Approaches and Resources* (Paris: NEA) https://oecd-nea.org/jcms/pl_14894/stakeholder-involvement-in-decision-making-a-short-guide-to-issues-approaches-and-resources

NEA 2017a *Bridging Gaps: Developing Sustainable Intergenerational Decision Making in Radioactive Waste Management* (Paris: NEA) https://oecd-nea.org/jcms/pl_19824/bridging-gaps-developing-sustainable-intergenerational-decision-making-in-radioactive-waste-management

NEA 2017b *Communication on the Safety Case for a Deep Geological Repository* (Paris: NEA) https://oecd-nea.org/jcms/pl_15032/communication-on-the-safety-case-for-a-deep-geological-repository

NEA 2020 *Preservation of Records, Knowledge and Memory (RK&M) Across Generations Final Report of the RK&M Initiative* (Paris: NEA) https://oecd.org/publications/preservation-of-records-knowledge-and-memory-rk-m-across-generations-50292bbb-en.htm

NEA 2021a *The Forum on Stakeholder Confidence Report on Dialogue in the Long-Term Management of Radioactive Waste* (Paris: NEA) https://oecd-nea.org/jcms/pl_56330/the-forum-on-stakeholder-confidence-report-on-dialogue-in-the-long-term-management-of-radioactive-waste

NEA 2021b *Managing Uncertainty in Siting and Implementation—Creating a Dialogue Between Science and Society* (Paris: NEA) https://oecd-nea.org/jcms/pl_59339/2nd-igsc-fsc-workshop-managing-uncertainty-in-siting-and-implementation-9-october-2019

NEA 2021c *FSC—Intergenerational Connections in Radioactive Waste Management: Involving Children and Youth* (Paris: NEA) www.oecd-nea.org/jcms/pl_61421/fsc-intergenerational-connections-in-radioactive-waste-management-involving-children-and-youth

NEA 2022a *Stakeholder Confidence in Radioactive Waste Management. An Annotated Glossary of Key Terms, Update* (Paris: NEA) https://www.oecd-nea.org/jcms/pl_63851/fsc-an-annotated-glossary-of-key-terms-2022-update

NEA 2022b *Added Value in Nuclear Waste Management Facilities and Processes* (Paris: NEA) www.oecd-nea.org/jcms/pl_63749/fsc-flyer-added-value-in-nuclear-waste-management-facilities-and-processes

NWMO (Nuclear Waste Management Organisation) 2021 Canadian Radioactive Waste Summit 2021 *Integrated Strategy on Nuclear Waste. What We Heard Report (1)* https://radwaste-planning.ca/sites/default/files/canadian_radioactive_waste_summit.what_we_heard_report_0.pdf

Reuters Institute 2019 *Digital News Report* https://digitalnewsreport.org/survey/2019/

Seaborn B 1998 *Panel Report* https://iaac-aeic.gc.ca/archives/pre-2003/431C8844-1/default_lang=En_n=0B83BD43-1_printfullpage=true.html#ws6B85477A

SFOE (Swiss Federal Office of Energy) 2006 Die Anwendung partizipativ*er Verfahren in der Entsorgung radioaktiver Abfälle* (Bern, Switzerland) https://pubdb.bfe.admin.ch/de/publication/download/1404.pdf

The Guardian 2020 *How Finland Starts Its Fight Against Fake News in Primary Schools* 29 January https://theguardian.com/world/2020/jan/28/fact-from-fiction-finlands-new-lessons-in-combating-fake-news

UNECE (United Nations Economic Commission for Europe) 1998 *Convention on Access to Information, Public Participation in Decision-Making and Access to Justice in Environmental Matters* (Aarhus: UNECE) https://unece.org/DAM/env/pp/documents/cep43e.pdf

Chapter 14

Ethical aspects of high-level nuclear waste management

Konrad Ott

14.1 Introduction

The management of high-level nuclear waste (HLW) falls into the category of collective action. Ultimately, states are responsible for its safe disposal, even if, as in Switzerland, companies are responsible in the first instance. Collective action differs from individual agency in some respects (Olson 1965), but they share the common idea of performing the 'right' (or 'best') course of action by making a reasonable choice between options. Ethics wishes to be helpful in determining such a choice.

HLW is *prima facie* a burdensome legacy for the future. '*Prima facie*' denotes that there might be exceptions. If these substances might be useful for some purpose in the future, they are possible assets. The HLW management debate focusses on the topic of how to dispose of these substances with or without prospects for future utilization. If one perceives HLW to be a burdensome legacy, the 'radwaste' debate is about finding the *least bad solution* for disposal. One may equate the 'least bad solution' with the 'best solution,' but such terminology downplays the fact that burdens, costs, dilemmas, or even evils remain (Kekes 1990). The 'best we can do' cannot nullify the burden completely.

Since HLW management is a 'wicked problem,' courses of action might be essentially contested. 'Wickedness' means that there is no commonly shared description of the problem, there is no algorithm for the solution, the underlying values, frames, and narratives are essentially contested, uncertainties (or even 'unknown unknowns') remain, and decisions have to be made (and are often urgent). 'Wicked problems' constitute what has been dubbed 'post-normal science' (Funtowicz and Ravetz 1993). 'Essentially contested' is borrowed from Gallie (1956) and adopts his conceptual idea for the analysis of wicked problems (see Ziegler 2020, pp 10–12). Gallie finally hopes for 'progressive competition' so as to reach the 'optimum fashion' of contested concepts. Thus, 'essentially contested' is not

doi:10.1088/978-0-7503-3095-4ch14

necessarily the final word on 'wickedness'. The contest over HLW management can be mediated and shaped, but also highlighted. Thus, contest implies different political profiles in different states. This chapter pays some attention to the German case, in which disagreement is high, but it does not take the German case as the paradigmatic case.

Practical philosophy reflects upon how to determine responsible courses of action by devices, procedures, principles, and criteria. These concepts are the main topics of this chapter. The next section takes some steps back into the history of nuclear energy (section 14.2). Section 14.3 introduces basic ethical concepts. Seven structural problems are identified in section 14.4. Section 14.5 offers some words on the themes of accessibility and reversibility. Section 14.6 analyses the concept of safety, while Section 14.7 addresses topics of justice. Section 14.8 provides some final reflections on building blocks and requirements for responsible HLW management.

14.2 The promise and fear of the nuclear age

The original imagery and promise of the so-called 'nuclear age' (Bluhm 1999) were of a presumptive solution to energy supply problems in advanced industrial societies. The growth rates of GDP were seen to be close to those of rising energy demands. Electricity was rightly seen as 'the' modern energy, and the provision of cheap electricity would spur economic growth and societal progress. Such beliefs had been dominant in both liberal-capitalist and state-socialist regimes. To many supporters of nuclear energy, protests came as a surprise. Protests intensified in Germany during the 1970s. The main reasons for the protesters' energy were: (a) proliferation and dual use, (b) catastrophic accidents, (c) disposal of radioactive substances and (d) threats to democracy. It seems fair to say that the cases of North Korea, Pakistan, and Iran confirmed the first argument. The accidents of Kyschtym, Three Mile Island, Chernobyl, and Fukushima may also count as confirmation of the second argument. In the aftermath of the Three Mile Island accident, Perrow (1984) wrote his book on 'normal accidents.' Argument (d), namely that nuclear energy might undermine liberty, rights, and democracy (Jungk 1977), has not been realized.

There was a moral discourse about HLW: is it ever acceptable to produce such substances? As Spaemann (2011) has argued since the 1970s, the answer should be 'No,' because such substances impose grave risks upon future generations. If *homo sapiens sapiens* has existed for roughly 250 000 years, one million years of safe disposal looks like 'mission impossible'—at least according to common-sense intuition. The counterargument pointed to geological disposal facilities that look solid over millions of years. Geology negates lay intuition. As the argument continues, future generations should be well-equipped with the capital and technologies required to find safe underground solutions. They might combine technological with geological barriers in order to achieve a high degree of safety. Future generations might even find new solutions that reduce time spans, such as transmutation. Although there were no real disposal facilities operating at the time, the outlook for safe geological deposits became a sufficient rationale for the building and processing of nuclear reactors in the 1970s. The history of HLW management in different countries is discussed in part II, chapter 12 of this book.

HLW management is not static, but evolves over time. It can have different contexts. It can be integral part of a nuclear industry that continues to operate (or even expands), or it can become part of the phasing out of such industries. In Germany, the political decision to phase out nuclear industries put management into a new context. Moreover, HLW can be associated with nuclear weaponry or civil industry. Therefore, ethical reflection is at the crossroads between generic topics and specific contexts. If it moves into generalities, it might collapse into truisms, while movement into specific contexts will be mired in details for which ethicists have no expertise. This chapter takes steps in both directions.

Recently, a new debate on nuclear energy has emerged in the context of climate change. If one believes that all humans are entitled to an affordable electricity supply, and if one regards renewable energies as insufficient to fulfill this entitlement completely, and if one wishes to keep climate change within a window of 'well below 2 °C' (or even 1.5 °C) increase in global mean temperatures (compared to pre-industrial levels), then nuclear energy becomes an option. The risks of catastrophic accidents might be reduced by a new generation of inherently safe reactors. Moreover, it has been argued that coal mining accidents have killed roughly 60 000 humans within the decade 1992–2002. The casualties of coal mining and the victims of climate change are far more numerous than the victims of nuclear accidents. But is 'counting lives' the right way to look at energy supply systems? The future of nuclear energy is open, but beyond the scope of this chapter.

14.3 Practical philosophy and ethical theories

Practical philosophy supposes that 'wicked problems' are to be deliberated via a broad pool of reasons. Reasons are linguistic units which are used as arguments within discourse (Toulmin 1958, Taylor 1961, Habermas 1981). In the case of wicked problems, there is no clear-cut hierarchy among reasons. Scientific, technological, economic, cultural, legal, political, and moral reasons must be deliberated, ordered, weighed, and, perhaps, negotiated. The broad landscape of reasons is loosely connected to the limited pool of options for disposal. Such connections can be made explicit in terms of 'if-then' reasoning.

Practical philosophy is broader than ethics, as it entails decision theory, political philosophy, theories of agency, legal studies, economics, and the logic of preferences. Ethics, as a specific field of practical philosophy, reflects upon moral validity claims. As this chapter shows, ethics must be contextualized within broader practical philosophy in the case of HLW management. Practical philosophy recognizes conceptual differences between morals and politics, discourse and negotiation, arguing and bargaining.

Generally, communication operates using the modes of *arguing* or *bargaining*. Arguing supposes that agreement is within reach, while bargaining recognizes that values and interests will remain different. Bargaining reaches out for compromise among parties, supposing that parties prefer compromise over conflict escalation.

Even discourse-oriented ethicists may say that in cases of wicked problems, a shift from discourse to (fair) bargaining might be more sincere than maintaining the illusion of discourse. Bargaining processes, however, rely on standards of fairness. *Negotiations* fall in between deliberation and bargaining. It is necessary to negotiate which HLW problems might be resolved via discourse and which ones must be handed over to bargaining. The rules of bargaining must be argued *and* negotiated. Policy making is an institutionalized combination of deliberation, negotiation, and bargaining.

As a mode of reflection, ethics analyses *moral validity claims* with respect to presuppositions, reasons, credibility, inferential structure, and implications. In ethics, arguing has a generic primacy over bargaining. We may bargain our contracts, but not our moral convictions. Moral validity claims refer to obligations and commitments. Commitments leave some leeway in the choice of how to fulfill them, while obligations determine a course of action that is mandatory. HLW management is committed to finding a 'safe' repository, but not obliged to dispose in specific ways.

Ethics always adopts some theoretical frame. A frame is a specific way: (a) to conceive moral principles, and (b) apply them to specific cases. Any frame has a theoretical core and a set of intended applications. There are different ethical theories, such as virtue ethics, contractarianism, utilitarianism, deontology, virtue ethics, just to mention the 'classical' ones.

Virtue ethics has deep roots in ancient philosophy (Aristotle, Stoic ethics, Confucianism). It identifies proper attitudes towards one's own life, close and distant people, institutions, and, perhaps, even natural beings. Such attitudes might be modesty, respect, compassion, fairness, gratitude, and humility. Virtue ethics defines the decency of individual persons, but it might be enlarged to the common decency of collectives. Decency rests upon habits and convictions about what humans owe each other in different spheres of life. MacIntyre (1981) argues that virtue ethics has been lost in modern times, but all ethical attempts to replace virtue ethics have failed. Following MacIntyre, one should not see virtues as obsolete. Virtue ethics suggest attitudes of trustworthiness, responsibility, decency, and prudence for HLW management, even if no strategy of disposal can be derived from such attitudes. Prudence should not be dismissed. Prudence means 'to be experienced' and 'to take an interest in good solutions.' Thus, prudence differs from egoistic cleverness. Recommendations suppose that they are given out of prudence. Deontology and virtue ethics can be combined if commitments and virtues are correlated.

To *contractarianism*, commitments emerge from contracts that free and reasonable persons agree upon. Some contractarians believe that valid contracts can emerge from thought experiments, while others demand 'real' contracts. There can be no real contracts between past, present, and future generations. Some contracts, however, might be tacitly implied in the chain of generations. If, for instance, a generation has received many benefits from former generations, such beneficiaries should not consume all the benefits and bequeath mostly burdens to their descendants. Such a strategy cannot be universalized and is, by intuition, unfair. The leading proponent of 'Kantian contractarianism,' John Rawls (1972, § 44), argues that rational persons, on being placed behind a veil of ignorance, would opt for a fair allocation schedule between generations as essential component of a fair political order. To Rawlsian variants of contractarianism, current generations are committed to bequeath fair legacies. There is an open question of whether a legacy can be fair in its entirety if it entails some burden. A person may accept private legacies if their overall values outweigh some debts, but will refuse legacies dominated by debts. Environmentalism argues that industrialized societies will bequeath too many burdens to future generations (Jonas 1979).

Contractarians may look for voluntarism in the case of repositories for HLW. If well-informed local communities deliberately decide to host a site for HLW disposal, if safety conditions are fulfilled and fair compensation is provided, this counts as a good solution to contractarians, since '*volenti non fit injuria*'. Any real contract between a local community and government would, however, have to guarantee that contracts between contemporaries are fair to future generations. If so, voluntary contracts are conditional on 'best' safety.

To *utilitarianism*, HLW management should maximize the difference between 'good' and 'bad' states from now to infinity, because utilitarianism implies a general commitment to maximize the overall difference between pleasure and pain ('good' and 'bad') via the consequences of courses of action. Welfare economics has utilitarian roots (Pigou 1952). The contemporary utilitarian frame entails concepts such as 'well-being', 'welfare', 'preferences', 'interests', 'harm', 'benefit,' etc. which ultimately refer back to mental states. To utilitarianism, the pleasure or pain (= harm) of all sentient beings counts morally. If people can live free from anxieties, this counts as pleasure. Threats affect the overall utility function, as fears are negative utilities. Intensities of pleasure and pain also matter. Since any unit of utility counts equally, regardless of its location, future utilities must not be discounted. Any utilitarian calculation, however, becomes imprecise (or speculative) if long-term uncertainties are at stake. If no precise utilitarian calculation can be made, utilitarianism is truncated to rule-based consequentialist assessment. Such an assessment is performed in any safety case (see part I, chapter 9).

The duty to maximize the difference between 'good' and 'bad' implies that short-term sacrifices are mandatory if they create more utility in the long run. If the overall sum of utility increases over time, no one can complain of being overburdened. If we were to burden the next few generations heavily to improve safety for the following 50 000 generations, utilitarianism would have to favor this solution. The supreme duty to maximize the difference between 'good' and 'bad' often clashes with both moral common sense and intuitions about fair distribution. Utilitarian sacrifices for the overall good can even permit killing.

Deontology wishes to substantiate a supreme moral principle, from which a set of generic obligations can be generated. The paradigmatic example of deontology is Kantianism, as the Categorical Imperative enables persons to check their personal rules ('maxims') according to the question of whether they might be suited to be the foundation of general societal laws. The Categorical Imperative improves on the Golden Rule ('Treat others only in the ways you wish to be treated by others') which is common in daily life and in many religions.

If one sees commitments and obligations as being derivatives of entitlements and rights, *right-based moralities* fall within the frame of deontology. If any right-bearer A has a right R, there must be some duty bearers B who have an obligation O to respect the performance of R or to fulfill R. The more rights are recognized, the more duties fall upon duty bearers. Rights can be violated, impaired, or compromised. Thus, there are strong and moderate denials of human rights. A slight impairment is morally different from a massive violation. A presumed 'right against polluters' (Kernohan 1995) must take different kinds of denial into account. There is no absolute right against any pollutant, but a civil right to responsible political risk regulation with respect to pollutants.

The full set of human rights has been expanded to material endowment rights (the right to food security, the right to fresh water, the right to health care services, etc). Some have argued that, in our digital world, there is a derived right to (affordable) energy, because without energy, in particular, electricity, other liberties and rights are no longer available (see Löfquist 2019 for an analysis). There is, as Löfquist argues, the danger of expanding the set of rights to anything that is beneficial to humans. Thus, the human right to cheap electricity can be rejected.

Deontology acknowledges obligations placed upon the members of future generations. If future persons are entitled to some rights, current persons have some commitments with respect to future rights (Unnerstall 1999). Detailed analyses are given in Düwell *et al* (2018). From the deontological perspective as outlined by Gewirth (1978), current generations must respect the conditions of agency and the liberties of future persons. Current generations are not allowed to overburden future generations in ways which impair such liberties.

Discourse ethics emerged from the framework of deontology. *Discourse ethics* combines the Kantian frame of supreme principles with the so-called 'linguistic turn' of modern philosophy. The linguistic turn shifts the basic question of philosophy from 'What are the basic structures of the world?' (cosmology) and 'How can a subject recognize an object?' (epistemology) to 'How can reasonable beings argue about matters properly?' In science, the linguistic turn analyses truth claims, while in ethics, moral validity claims are checked (criticized, justified). Discourse ethics substantiates some general commitments of argument that are constitutive to the practice of reasoning. Such rules of discourse cannot be denied by participants in discourse without self-contradiction (Ott 2017, following Apel 1976 and Habermas 1981). The principle of discourse ethics, as derived from the rules of discourse, concludes that moral commitments (rules, duties, institutions, rights) are valid only if they can (might) be agreed upon by all participants in discourse. This supreme principle must be applied to moral rules and legal institutions (Habermas 1992).

Discourse ethics is procedural and, by definition, discourses are open. Only the force of better reasons should count in discourse. The forces of such reasons cannot be observed from outside, but must be revealed to participants from within. No single participant can know the results in advance, because reasons must circulate among a plurality of participants. This is *a fortiori* true in cases of wicked problems.

To discourse ethics, there are some grains of moral truth in other ethical theories. If so, there can be virtue-based, contractarian, consequential, and deontological reasons within the entire pool of reasons in moral debates. In the cases of future generations, for fetuses, mentally handicapped humans, and other 'moral patients' there must be advocacy in discourse.

Discourse ethics distinguishes 'pure' moral discourse from legal and political debates. In political debates, the pool of reasons enlarges and includes non-moral reasons. In practical decision-making, moral reasons have to compete with non-moral reasons. Economic opportunity costs, cultural traditions, technological feasibilities, constitutional law, federal structures, etc. cannot be neutralized. In political debates, it becomes questionable whether moral reasons always trump ('override') all other reasons. If moral reasons can always override all other reasons, policy making becomes 'applied moral philosophy.' Moral concerns about repugnant side-effects (or 'slippery slopes') do not override all other reasons.

The enlargement of discourse ethics into policy making has been conceived as deliberative democracy (Habermas 1992). Deliberative democracy rests on the two pillars of inclusion and deliberation. 'Inclusion' means that all persons affected by regulations should have a voice. 'Deliberation' is about political argument within institutional settings. Deliberative democracy distinguishes (a) the core of democratic policy making (parliament, government, supreme court) from (b) civil society (NGOs, parties, pressure groups) and (c) the middle ground of boards, offices, and settings (such as boards for scientific policy counselling) (Habermas 1992, chapter VIII). Collective decision-making is restricted to institutions with input legitimacy, such as parliaments. Institutional governance is located at the interface of (a) and (c). Decisions should be prepared via deliberation. Deliberative democracy is sympathetic to participatory formats, but it should not be overly idealistic with respect to participation in cases of wicked problems, as we shall see in the next section.

14.4 Structural complexities

If deliberative democracy is applied to HLW management, some structural problems (complexities, challenges) become evident.

First, there is an enlarging pool of reasons. Prescriptive reasons must fully recognize the epistemic side of the debate, such as geology, radiology, technology, etc. Epistemic and normative premises, taken together, form a practical judgement to which reasonable agents might agree. It remains possible to agree on normative principles, but disagree about empirical evidence, modelling, etc. All reasons must pass a threshold of relevance. There might be disagreement about whether the outcomes of a specific scientific model pass such a threshold of relevance.

Second, deliberative democracy demands inclusion in terms of participation. Participation includes engaged citizens in deliberative settings, as in so-called citizen juries. Meanwhile, some participatory formats have been integrated into HLW management in all European states. There is a tendency toward inclusiveness, participation, mutual learning, open debate, consultation, etc. in HLW management. Since there is no established metric for participation, citizens and pressure groups may see such settings as not 'fully' participatory, while governance agencies may claim that the setting is 'fair enough.' In the German case, such agencies have been accused by participants and some NGO's of not reaching sufficient levels of participation (see the analyses in Themann *et al* 2021a, 2021b). Such accusations are grounded in standards for appropriate levels of participation. Standards can be either moderate or demanding. Thus, discourse ethics should reflect the background standards for adequate participation. In principle, participation can be either (a) insufficient, sufficient, or ambitious (in terms of quantity), and (b) it can succeed or fail (in terms of results). Both distinctions allow for nuanced judgement. One may, for instance, argue that (a) a format was successful, even if it was insufficient, and (b) that ambitious formats have failed (for whatever reasons).

A *third* point deserves attention. As sociologist Niklas Luhmann pointed out (1991), the search for safety based on calculations and estimates of risk will always give rise to concerns that risks (threats, dangers) are imposed upon persons without consent. Risks are estimated by scientifically trained experts (as in safety cases, see part I, chapter 9) but the results of such estimates may be perceived as (horrible) threats by affected persons. The difference between experts (estimation) and civil society (anxieties) makes a political difference in terms of protest. Bechmann (1993) argued that discourse ethics is bound to fail in cases of risk and uncertainty in which concerns, distrust, and anxieties are dominant. Even if all persons were rational, different degrees of risk aversion would result in disagreement. To Luhmann, the gap between the societal positions of risk–decision and threat–concern cannot be bridged by improved participation.

Fourth, trust and distrust play a decisive role in HLW management, but they vary among nations. If distrust prevails, scientific estimates will not temper emotional reactions. Trust and distrust fall within the scope of virtue ethics, as they are attitudes toward other persons, authorities, and institutions. There are widespread beliefs that distrust is more 'critical' than trust. There is, however, also the attitude of 'uncritical distrust.' Trustworthiness among participants is an essential condition for discourse, while common deliberation may enhance trustworthiness.

Fifth, perceived threats and an attitude of distrust generate emotions such as anger, fear, or even rage. In our culture, anxieties are widespread (Nussbaum 2018). Ethics faces ambiguities with respect to emotions. On the one hand, ethics must make room for proper emotions (such as compassion), since moral affairs cannot be perceived in cold blood, while, on the other hand, affections and passions should not trump scientific, technological, and economic reasons. In contested matters of risk perception and regulation, there are no clear criteria for 'appropriate' emotions. This holds true for HLW management. Mixtures of distrust and anxiety may pose huge challenges to deliberative and participatory processes. Participatory formats should not take emotions as given, but should motivate participants to reflect on them.

Sixth, there is high likelihood of rhetorical and strategic reasoning within political debates. Moreover, there are *theories* of discourse that see discourses as battlefields governed by power, rhetoric, strategy, tactics, and persuasion (Foucault 1973, 1984). If one such adopts agonistic approaches to discourse, one will feel entitled to make speech acts strategically without bad conscience. Agonistic theories of politics see deliberative democracy as a liberal illusion (Mouffe 2005). It remains doubtful whether agonistic approaches can address the HLW case properly. These theories will not work well in Scandinavia and Switzerland, and they cannot be applied to HLW management in authoritarian regimes. An interesting article combined Foucault's theory of discourse with Fish's juridical–political communication theory and applied both approaches to the Yucca Mountain HLW management project (Lipscomb and O'Connor 2002).

Seventh, HLW management refers to topics of *justice*. Rawls, in his '*Theory of Justice*' (1972), draws a distinction between moral duties (Rawls: 'natural duties') and the principles of political justice. Justice is, as Rawls argues, the foremost qualification of political institutions. It remains close to morals, but it also includes politics, law, and economics. Rawlsian justice refers to the basic structure of a fully liberal democracy, being defined as a 'well-ordered society'. To Rawls, 'well ordered' is different from 'law and order' because the order is grounded in sound principles of justice. In the following, HLW management in authoritarian regimes is abstracted away.

A fully liberal democracy is free to decide about its energy supply systems. Thus, a decision in favor of nuclear energy is legitimate from a Rawlsian perspective, and so are decisions to establish, phase out, or maintain it. Fully liberal democracies are, in general, just, but there can be many instances of unfairness in the details. Fully liberal democracies gradually enhance the standards of fairness in different spheres of life: education, gender equality, inclusion, health care, and even the economy. They also expand the electorate stepwise. Fully liberal democracies have to distribute benefits and burdens. Distributions will affect persons unevenly. Noise, pollutants, dust, highways, waste management, hard labor, and constraints on the one hand, but also benefits, access, transfers, support, etc. on the other hand are dispersed unevenly across space, times, and groups. Therefore, there will be many opportunities to make a matter of justice out of distributions.

There are two approaches to justice. The ideal-justice approach defines an 'ideally fair' solution and wishes to bridge the gap between ideal solutions and non-ideal realities (see Carens 2013, pp 300–303 on the 'just world presupposition'). The satisfaction approach defines thresholds above which solutions are 'fair enough.' The difference is between idealism and pragmatism. Discourse ethics is not committed to idealism in any case. With respect to HLW management, seen as a wicked problem, a pragmatic approach looks more appropriate. If wicked problems have no perfect solution, it may be incoherent to combine them with ideal-justice approaches. Such a combination may perpetuate wickedness. Wicked communication may emerge if some people suppose ideal justice while others just suppose satisfactory solutions. HLW management is not committed to finding 'ideally fair' solutions.

To wrap up this section: since the pool of reasons expands in political debates about wicked problems, since different ethical approaches are supposed, and because agents make use of rhetoric with strategic intentions, the likelihood of perfect agreement is low even if the level of participation is high. HLW debates reveal a peculiar dialectics in between the suppositions and the realities of argument. While it must be supposed that agents argue and, by doing so, reach out for mutual understanding and agreement, the multitude of reasons and stakes imply disagreement. Discourse ethics and deliberative democracy expect consensus only if regulations are impartial, i.e. in the common interest of all persons being affected. This condition of impartiality is unlikely in cases in which hosting a repository is primarily seen as a burden to some. In such cases, deliberative democracy points toward fair institutional procedures that include negotiation and bargaining.

The German Site Selection Act (StandAG 2017) provides an example. As § 5 states, there should be a broad societal agreement about a solution which can be tolerated by the groups that are negatively affected. This legal statement implies that the 'broad agreement' will be less than perfect. The distinction between 'broad agreement' and 'toleration' implies that the affected groups may not agree. On a second look, § 5 wishes to avoid an escalation of conflict. Broad agreement would mean that politics remains in control, even if some acts of protest and civil disobedience occur.

Discourse ethics is on the boundary between counterfactual ideals and illusions. Counterfactual ideals are enabling conditions of discourse, while illusions may overlook the sober realities of political debate. Illusions of discourse should be replaced by realistic outlooks: anyone can be expected to argue in the search for the common good but everyone wishes to succeed in their own interest. This dialectic results in 'wicked communication.' If so, discourse ethics has a threefold task: first, it must hold the standards for deliberation high according to commitments made in arguments and collective epistemic memory established over decades in HLW communities worldwide. Second, it must identify kinds and instances of wicked communication. Third, it must point to the limits of discourse. Beyond the limits of discourse, there might be decent ways to deal with the problem of allocating HLW by means of negotiation and bargaining. There might be an 'abstract' consensus ('safe and fair disposal') which stands in need of bargaining to determine the details.

14.5 Accessibility and reversibility

If the concept of deliberate democracy can, despite all its complexities, be applied to HLW management, the full set of options must be taken into account. This set is, of course, limited. Proposals for disposal strategies are (a) deep geological disposal (1) with or (2) without access and (3) with or (4) without options for retrieval, (b) intermediate surface storage, (c) rolling stewardship, and (d) 'exotic' options (outer space, deep sea, core of planet, etc). The so-called 'exotic solutions' have been rejected by most experts and are abstracted away. The case for surface storage was made by Shrader-Frechette (1993). The argument was science-based. Surface storage should buy time to search for the best available geological formation. 'Rolling stewardship,' as proposed by Gordon Edwards in many articles, pleads for surface storage without end. It puts burdens and risks upon future generations, since it requires periodic repackaging and continued monitoring. Ethically, it is based on communitarian ideas of how local communities should take care for HLW. It is also based on ideas of the persistence of memory. A long-term solution, however, should not rest on communitarian ideas, because local communities are fragile and may disperse. Thus, 'rolling stewardship' options should be rejected.

An analysis of surface storage is given in Ott and Budelmann (2017). As Ott and Budelmann argue, one can upgrade storage buildings to improve technological safety, but such buildings remain an intermediate solution, which implies the risk that policy makers may become disinterested in the search for a final repository (see the contributions in Köhnke *et al* 2017). Surface storage has deficits in terms of safety and security. Therefore, the overall strategy should terminate in a geological repository as the 'end point.' Such a repository can either be accessible or, in practical terms, inaccessible.

The options for accessibility, retrievability, and reversibility became transcendent topics in the German case. All three concepts are dispositional terms, which implies that the dispositions may become actual: 'access,' 'retrieval,' and 'revision.' We have to debate prescriptive dispositional concepts in claims such as: 'It should be possible to do *x*.' The open question always remains: 'Why?' The answer can either be: 'Because such a possibility is good in itself' or 'Because such a possibility is good in specific cases.' In the case of access, generally speaking, access to weapons, child prostitution, and drugs is not good in itself. Thus, one must be more specific in the case of HLW.

Accessibility can refer to the option of utilizing the substances (transmutation, fuels for new types of reactor) which can be justified by the presumed higher liberties of future generations. Accessibility can also be favored in order to correct failures and accidents (for example, the leakage of toxic substances). While the former means accessibility, the latter means retrievability. These meanings are often confused. Retrievability only applies in the case of accidents. It is possible to make a repository rather inaccessible but leave some options for retrievability in case of accidents. A safety case may favor a combination of inaccessibility and retrievability, while policy makers may favor reversibility as a more charming concept. If one wishes to have both inaccessibility and retrievability, one must weigh the details of combinations which are beyond the expertise of ethicists.

Access can be given for specific periods or indefinitely. As has been argued elsewhere (Ott 2020), there is no obligation to provide access indefinitely. Any generation is entitled to make access practically impossible. In some 'classical' strategies, accessibility decreases over time. It can be morally right to deny access in specific cases. The claim 'Person P should not have access to X' can, in principle, be sound.

Reversibility is no value in itself. There are many former practices whose abolition should remain irreversible for moral reasons, such as slavery. Thus, there remains an open question of whether some specific X should be regarded as, or made, irreversible. Reversibility can refer to the selection process of a disposal site. If procedural justice requires learning and corrections, there must be some leeway for revision. The revision of crucial decisions made in HLW management, however, comes at high political and economic cost. The truth in claims for reversibility is to pay attention if one reaches a point of bifurcation. Such points are the 'points of hard returns.' At these points, participation using well-established formats (such as citizen juries) makes perfectly good sense. Finally, however, there must be a decision by an authority which has high democratic legitimacy (parliament). To propose a revision of a crucial decision on the pathway to a final repository, then, implies a burden of proof. The more steps on the route have already been taken, the more costly any revision becomes. Imagine, for instance, a scenario in which a decision is made in favor of salt-based geology. A site has been selected, a repository has been built, the substances have been handled for deposition, but then some agents propose a clay-based geology based on safety concerns. Might, even if the concerns are reasonable, a government argue that the geology case has been closed for decades? Can there be valid arguments against demands for revision that state 'It's simply too late *now*'? Other limits of reversibility include aging containers that reduce safe handling and reconditioning for a different host rock.

Accessibility and reversibility will (for technical reasons), and perhaps should, decrease over time. HLW management might intend such a decrease. If so, management cannot maximize options but has to reduce them. It must determine a final solution and reach out for an 'end point'. Future liberties are, in principle, compatible with the reduction of some options. There is no moral principle that requires a continual increase in the number of options. A mere increase in the number of options will not make the world a better place.

14.6 Safety

The value of safety as such is beyond dispute. One must be more specific: 'How safe is safe enough?' *Safety* in a broad sense has two dimensions: (a) technological safety and (b) security against misuse. Safety cases can be made for a type of repository or a token ('site'). Safety in a broad sense can be assessed by safety cases (see part I, chapter 9 of this book). A distinction must be made between perfect and sufficient safety. Since perfect safety is an ideal ('zero risk'), safety must be vaguely defined as the 'best available'. The maxim 'to reach out for the best' is, however, always constrained by limitations of time, knowledge, and resources. 'Best' is located somewhere in between ideal and sufficient safety: more than merely sufficient and less than perfect.

With respect to a site, expressions such as the 'best available safety' can either refer to the 'very best solution' (= better than all other solutions) or to a set of 'best' solutions compared to 'less good' solutions. This set, however, can entail more than one element. This reading of 'best' is similar to the use of 'best' in utterances such as: 'She is one of the best piano players worldwide.' Since safety cases based on models cannot identify small degrees of difference in safety over long time horizons with certainty, the best they can do is to identify a set of 'best' solutions with ($\in > 1$). If so, the decision for a specific site will be a political one, informed by science, but not completely determined by science.

The search for the 'best' solution requires a selection process within which the number of options decreases. In such selections, comparisons play a crucial role. Criteria operate within comparisons. Sufficiently safe is not good enough, if there are still betterness relations to be determined. 'Best' is reached if comparative safety cases no longer bring clear results in terms of betterness relations. The relation 'A better than B' transforms into the relation 'indifference between A and B'. If there is more than one 'best' solution, non-safety-based considerations may determine the final decision (efficiency, voluntariness). If two sites, A and B, cannot be distinguished in terms of safety, but B is far cheaper to build than A, efficiency demands that B should be built (and the money saved spent on another 'good' project).

The 'minimax criterion' might be helpful in defining a safety threshold. It requires that the worst possible (conceivable) outcome will not be catastrophic. If society will always have to cope with extreme events, a repository will be 'safe enough' if the 'worst-case' damage and casualties cannot reach (with high likelihood) even a substantial fraction of those of such extreme events (earthquakes, floods, hurricanes). The minimax criterion with respect to safety and security makes an argument against surface storage, as the worst case might be worse at the surface. Geological barriers will continue to function as protective zones against release into the open environment.

With respect to *types* of repository, both technological safety and security are higher in deep (and practically inaccessible) geological formations than in surface storage. Since long-term societal change is unpredictable, accessibility implies a deficit in security. Climate change, migration, social unrest, political extremism, etc. might take place. Accessibility supposes 'good' intentions, but it cannot rule out a desire for access caused by evil ones. Although faith in progress has been shattered over the last century (Horkheimer and Adorno 1947, Jonas 1979), some surface, access, reversibility, and stewardship options seem to be based on a faith that the future world will allow for permanent HLW management. Practically inaccessible underground disposal may liberate future generations from one specific problem, since they will have to cope with many others, for example, climate change, pandemics, the conservation of nature, irregular migration, and even war.

The (often underrated) dimension of *security* has to address how a surface storage facility might be defended if it were to come under attack. Defense strategies would have to include both military and psychological resistance, as they must expect violent and cruel means. Unwillingness to talk about defense might count as a concession in favor of practically inaccessible underground storage.

Safety clearly has a long-term perspective (up to one million years) (Kermich 2015). The future, however, is not homogenous (Ott and Semper 2017). There might be a trade-off between near and remote posterity. If a society takes more time to research presumptive geological sites (salt, clay, crystalline), it takes more surface storage risks over the forthcoming two or three generations. Should one, with certainty, burden the generations of one's children and great-grandchildren in order to increase, perhaps, the degree of long-term safety a bit? Here, we rush into the fuzzy logic of 'a bit more or less'. What are the burdens and costs that may buy 'a bit of additional safety' (under a minimax criterion)?

If a society wishes to speed up HLW disposal without delay, it may do so at the expense of remote future generations, since not all presumptive sites will be explored in line with due diligence. The trade-off is relevant in terms of societal opportunity costs. It is also relevant in terms of participation, since speed might reduce the time needed for 'genuine' participation. Deliberation takes time and argument takes place in slow motion. The trade-off in timeliness also affects whether the technological expertise ('human knowledge') will continue over time, may increase, or will fade away. This multi-dimensional trade-off with respect to speed and time has no common measure. It is incommensurate and must be prudently weighed. If HLW management will burden future generations until the end of the 21st century, we would do better by not overburdening the chain of generations with a 'never-ending search process'. In some sense, contemporary elderly adults are the beneficiaries of the wealth-generating industrial economy of the post-war decades, which was based on fossil fuels and nuclear energy. Beneficiaries who wish to act responsibly for the near-term future should, as I wish to argue, accept some opportunity costs rather than delay. We should take action by ourselves, must burden our descendants, and should try to liberate the remote future.

Opportunity costs mean that any option comes at a price. In economic parlance: the realization of the first preference costs the second-best. If P has a million euros and she has to choose between a Picasso or a mansion, the choice of the mansion will 'cost' her the Picasso. If a society allocates some additional billions to HLW management, the opportunity costs may affect health care, education, tax release, subsidies for organic farming, etc. As long as there is scarcity, there will be opportunity costs. The criterion of efficiency is not opposed to safety, but it has to balance degrees of additional safety and opportunity costs. Imagine that a container made of gold will increase long-term safety a bit but will cost some billions of euros. HLW management cannot and should not abstract away societal opportunity costs. The wording of 'best available' safety should not camouflage the fact that HLW management has to deal with opportunity costs. An open political debate about opportunity costs is rather unlikely, because many people will argue that 'safety should not be for sale.' Such reasoning, however, is superficial moralizing. Even if we believe that human dignity is beyond price, there will be decisions that may affect statistical human lives, as in the fields of medicine and traffic. The problem of a future statistical life unites economics and ethics in theory, even if these disciplines divorce in terms of results.

Efficiency criteria and opportunity costs are sensitive to the *rate of discounting* (see the contributions in Hampicke and Ott 2003). It has been argued that the rate of discount should be zero in HLW disposal (Schulze *et al* 1981). If one focuses on toxicity as such, a zero rate seems adequate. From a theoretical perspective, however, rates of discount should vary according to society's capability to cope with problems. If the future capacity to treat cancer or deal with toxic substances is seen as higher than today's, one may argue for a positive rate of discount. If the rate of discount is positive, however, one can be forgetful about remote accidents in deep underground repositories (under a minimax criterion).

14.7 Justice as fairness

The dimension of *justice* is crucial for HLW management. It has an inter- and intragenerational dimension. The former is covered by the 'best' safety dimension under the minimax criterion, sympathies for inaccessible repositories, high value for future statistical lives, and a low rate of discount (see the previous section).

Approaches to intragenerational justice can be either idealistic or pragmatic. With respect to HLW management, there are no firm standards about an ideally (= perfectly) fair procedure within democratic states. Political fairness will remain imperfect. If so, any fully liberal democracy must determine standards of sufficient fairness. The question 'How safe is safe enough?' has an analogy in the field of justice: 'How fair is fair enough?'

As we have known since Aristotle, there are different realms of justice. One can distinguish (a) procedural, (b) distributive, and (c) compensatory justice. The dimension of recognitional justice underlies all dimensions. Recognitional justice asks which kinds of recognition we owe to each other in specific spheres of societal life. It must be specified in terms of gender, education, status, merit, age, profession, and authority. This sophisticated dimension of justice looks rather irrelevant in HLW management. Some may add restitutive justice, in which former enemies wish to make new beginnings by way of reconciliation after severe cases of injustice (such as mass atrocities). Restitutive justice is abstracted away, since it requires conciliation tribunals, confessions of guilt, and pleas for forgiveness. Who among the living would have to declare herself 'guilty' for HLW at such a tribunal?

Democracies incorporate a tendency for the ongoing democratization of societal life. Demands for 'more democracy' intensify within democracies. Democratized HLW management turns into *procedural* justice. The entire search process (including the type of repository and the selection of a site) should be procedurally fair, the criteria for which are transparency, credibility, and participation. Thus, HLW management is forced to include broad participation. In Germany, participation is required by law. The 'utopia' of participation hopes for common learning processes by which state and non-state agents unite in the search for the 'best' solution. Degree, modes, and formats of participation may vary in between countries. Deliberative democracy requires participation to improve the deliberative processes by which political decisions are prepared. Participation should be inclusive and deliberative, but it demands neither voting rights nor veto rights for specific groups. High degrees of participation do not imply that all parties and stakeholders will 'win.' Participation improves the legitimacy of democratic decision-making, but a presumed deficiency in 'real' participation does not render a democratic decision illegitimate. The problem remains that deficits in participation are easy for NGOs to claim but hard for governance agencies to refute. Part II, chapter 13 in this book compares participation in different countries.

Procedural fairness is not identical to specific levels of participation. One may ask, for instance, whether a lottery for a repository site might bring about procedural fairness in cases of equal safety. Lotteries are instances of pure procedural fairness that decide cases of undecidability. If lotteries are agreed upon and are not manipulated, the outcomes are fair.

Fair procedures generally increase the likelihood of good outcomes. A good outcome in general might, however, be a bad outcome for someone. Even if the selection process is sufficiently fair, the inhabitants of the selected repository site may emotionally feel like losers. Even if they cannot complain about unfair process, there is likelihood that they will struggle against the result. Even the 'least evil' is an evil for someone. We can, however, modify degrees of hardship and pay compensation, even if the procedure was fair enough.

Distributional justice is about how to distribute goods and/or bads. HLW management means, *prima facie,* the allocation of negative good. In the case of negative goods, we intuitively wish to distribute burdens onto many 'shoulders.' Such a 'dispersal' solution, however, conflicts with safety criteria and economic costs. To disperse HLW multiplies risks. Thus, disperse-and-dilute-strategies are rather repugnant, and a pooled negative good must be allocated. The '*prima facie*' indicates that there might be different cases. If a local community sees opportunities for development in becoming a site for an HLW repository, some people may lament that such communities will often be poor communities which suffer from general inequality in the distribution of national wealth. If so, voluntary decisions will be criticized in terms of so-called 'background injustice.' It remains, however, open whether the differences in wealth regions are the outcomes of distributional injustice or regional predicaments and matters of historical and geographical fortune. In arguments from background injustice, the allocation of HLW is judged against a backdrop of many allocations of benefits and burdens. The paradigmatic cases are patterns of allocation which disadvantage marginal groups, as in cases of 'environmental racism.' Apart from clear-cut cases, holistic approaches that evaluate specific allocations against the entire distributive horizon open endless controversies. Discourse ethics rather warns against starting political debates that are without end.

In principle, the voluntary adoption of HLW remains an option for local communities if and only if the location is among the 'best' from a safety-case perspective. The political wish to circumvent a decision by voluntarism should not come at the expense of safety. Under this requirement, any local community might, in an uncoerced and well-informed manner, decide to become the location for the repository site, as it happened in Sweden (see part II, chapter 15). If and only if the criteria of safety will not reduce the number of 'best' sites to one, voluntariness may play the role of a 'tie-breaker' among the remaining candidates. Given different conditions, there might be both honeymoons and divorces between safety-driven approaches and voluntary approaches in different states.

As Ott and Riemann (2018) argue, there are two variants of voluntariness. The strong variant offers a sacrifice for the common good, while the weak variant includes bargaining for compensation. Since democracies do not usually demand sacrifices, the weak variant is a better fit for our political and economic culture. Compensatory practices are common, but not all burdens and disadvantages qualify for compensation. If the price of real estate decreases near housing for refugees, no entitlement for compensation is provided. Compensation requires victimization as a kind of wrongdoing.

The foremost question is whether compensation must be paid at all if the selection procedure has been sufficiently fair. Procedural justice might make compensation unnecessary. One may argue that fair procedures do not victimize. If the allocation is burdensome but not an instance of victimization, compensatory payment might nevertheless be politically prudent. If one sees allocation as (close to) victimization, compensation becomes mandatory. The moral contest over victimization might be reduced to the theoretical question of whether compensation should be given out of duty or out of prudence.

If the preceding question is answered in the affirmative, the second question is how to avoid both under- and overcompensation. There might be a politically induced asymmetry in favor of overcompensation. Since politics will be freed from making a contested decision in parliament or by government, politicians might abandon the economic criterion not to over-compensate. If the overall citizenry will be also glad, politics can be generous. Given the lump sums involved in HLW management, generosity is not costly, even if the opportunity costs of overcompensation cannot be denied. Given the prospect of prudent generosity, local communities may even compete for the status of repository site, as they hope to bargain for generous compensation.

If compensation should be offered, different models for compensation can be offered. Compensation models should be conceived in participatory ways. In participatory formats, people might not just rely on actual preferences, but might also mobilize imagination and fantasy (see Weston 2012). Participatory formats may transcend conventional models of compensation.

The creation of jobs due to the construction of the repository should not count as compensation. The state might, however, relocate some organizations and agencies whose clerks bring high purchase power to the affected region. The 'classical' model would compensate by improving infrastructure. As infrastructures have improved in all Western countries, this model looks outdated, since additional infrastructure provides fewer incentives. New *'green'* models of compensation might include subsidies for public transport, ecological forestry, organic agriculture, 'green' adaptation to climate change, restoration, public parks, etc. *Liberal–economic* models may include tax relief, basic income, reduced pension age, higher minimum wages, additional holidays, health care vacations, etc. *Post-modern* models might also include animal sanctuaries, full gender equality, post-colonial inclusion, queer artistic life, subsidies for festivals, legal marijuana, etc. Whatever one might favor, compensation models should think 'outside the box' of conventional regional development. Note that all models need to limit entitlements in space and time and will set incentives to migrate into or out of the site location. Compensation might make the site location attractive and even increase the economic value of real estate at the repository site. One should not suppose that current anxieties will remain constant over time. Future people may feel safe at the site and, in daily life, they might become forgetful about the repository, even if technological knowledge about the architecture, barrier structures, and contents of the repository must be kept alive. If behavior reveals preferences, a criterion for successful HLW management is an increase in population close to the site. If so, HLW managers should act accordingly. The maxim should be to make the site 'attractive.'

14.8 Conclusions

Irrespective of whether it was wrong or permissible to produce highly radioactive substances, they exist, and we cannot wish them away. The guiding question is this: How can a world with HLW come as close as possible to a world without it? The existence of HLW should be 'minimized.' Minimizing does not mean 'hiding,' rather, it means 'neutralizing.' The initial appeal of 'outer-space' solutions rest on the very idea of removing HLW from 'our' world, even if risk assessments reject this option. The underlying concept of *world* deserves attention. To Ludwig Wittgenstein, the world is everything that is the case. It is the entirety (totality) of facts, not of things (see, for interpretation, Stenius 1960, pp 18–20). Nothing exists outside the world. HLW will remain both as facts and things within the world. The concept of a world, however, has a different meaning in ordinary language. It is the 'world that surrounds us,' also known as our 'everyday world.' Management will keep HLW within the world, but also remove it from the everyday world. One may speak of 'abstract neutral existence.' In the longer run, safe disposal will reduce existence to knowledge about existence.

Humans live on the terrestrial surface of planet Earth. This may count as an anthropological argument in favor of deep geological repositories which are practically inaccessible. The fact that such 'end points' were conceived by the 'pro-nuke' communities does not count as a refutation of the very idea. If so, former opponents can agree on such an end point, which would finally allow for gradual policy termination if these substances will not be produced subsequently.

Neutralizing HLW should not make the site invisible. Future generations must be aware of the 'where, what, and how' of a geological repository. This is the topic of knowledge to be continued. Since humans are linguistic beings and can interpret signs, contemporaries might consider how to conceive a 'legacy of knowledge,' including warnings ('hands-off' signals). Since humans are curious beings, such warnings may, however, attract them. A perverse attraction to the 'forbidden' is *prima facie* beyond the scope of moral responsibility. Since humans are, finally, symbolic animals, one might even design the surface constructions of the geological repository as an architectural sculpture, perhaps megalithic.

Within the framework of discourse ethics and deliberative democracy, we are finally able to identify the crucial building blocks and requirements of reasonable and morally decent HLW management:

- Ethics will not determine the 'right' solution, but identifies the requirements of 'fair' processes. If ethicists make substantial claims, they take the role of engaged citizens.
- Deontology creates a commitment to dispose of HLW safely. Ethics analyses the meaning of 'best,' but cannot identify the best.
- From a consequential assessment, safety can be specified according to a minimax criterion. The combination of geological and technological barriers should specify this criterion.
- HLW management is a wicked problem, full of complexities that generate wicked communication. Ethics has to analyze utterances with respect to

wicked communication. Deliberative democracy demands that wicked communication should be combated within deliberation via ethical analysis.

- Discourse and deliberation must be augmented by negotiations and bargaining processes.
- The common decency of the epistemic community is a necessary requirement for societal trust. Trust must be deserved by performance within HLW management. If such decency can be supposed by the broader public, the burden of proof falls upon distrust.
- HLW management must be contextualized into overall prospects for the future. It should reduce the number of future problems and, by doing so, liberate future generations.
- There is a trade-off between the close and the remote future. A slight increase in long-term safety must be weighed against the burdens imposed on the next few generations.
- Deliberative democracy and procedural justice require adequate standards of public participation in order to find reasonable strategies and prepare decision-making. Participation cannot replace democratic decision-making.
- HLW should not be dispersed and diluted.
- Reflections on accessibility, retrievability, and reversibility speak in favor of deep geological and practically inaccessible repositories as end points. Retrievability specifies the minimax criterion.
- Security should not be underrated, since fanaticism, terrorism, (civil) wars, etc. are part of the human predicament. One should not build repositories on hopes for eternal peace among humans.
- From contractarianism, one can adopt the search for voluntary solutions and adequate compensation. Management should invest some fantasy in compensation models. It should make sites attractive.
- The voluntary agreement of the living excludes posterity. Therefore, voluntariness should suppose the 'best' long-term safety and should not come at the expense of safety from an ethical perspective.
- Finally, democratic decision-making and the rule of law are unavoidable. Fully liberal democracies can live with protest and civil disobedience.

These building blocks remain open for further debates. They can be varied in different nations since they only provide orientation at best. Orientation does not determine details. Discourse ethics would be self-contradictory if it were to prescribe any detail of HLW management. The interplay between safety and voluntariness will shape HLW management in different countries. The open-ended narratives of single-country studies of how this is done might be checked against these requirements. A hypothesis can be derived from ethical reflection: if country-specific management schemes are judged, the fulfillment of requirements should show a 'smooth' political process. If one can identify success stories, it may stimulate management in other countries.

Bibliography

Apel K-O 1976 *Transformation der Philosophie* (Frankfurt/M: Suhrkamp)

Bechmann G 1993 Risiko als Schlüsselkategorie der Gesellschaftstheorie *Risiko und Gesellschaft* ed G Bechmann (Opladen: Westdeutscher Verlag) pp 237–76

Bluhm H 1999 Das Atomzeitalter: Varianten einer Epochenbestimmung *Neustart des Weltlaufs* ed K Fischer (Frankfurt/M: Suhrkamp) pp 203–24

Carens J 2013 *The Ethics of Immigration* (Oxford: Oxford University Press)

Düwell M, Bos G and Steenbergen N V (ed) 2018 *Towards the Ethics of a Green Future* (London: Routledge)

Foucault M 1973 *Die Ordnung des Diskurses* (Berlin: Ullstein)

Foucault M 1984 *Der Foucault Reader* ed P Rabinow (New York: Pantheon Books)

Funtowicz S and Ravetz J 1993 The emergence of post-normal science *Science, Politics and Morality* ed R V Schomberg (Dordrecht: Kluwer) pp 85–123

Gallie W 1956 Essentially contested concepts *Proc. Aristotelian Society* vol 56 pp 167–98

Gewirth A 1978 *Reason and Morality* (Chicago: University of Chicago Press)

Habermas J 1981 *Theorie des kommunikativen Handelns* (Frankfurt/M: Suhrkamp)

Habermas J 1992 *Faktizität und Geltung* (Frankfurt/M: Suhrkamp)

Hampicke U and Ott K 2003 Reflections on discounting *Int. J. Sustain. Dev.* **6** 7–24

Horkheimer M and Adorno T W 1947 *Dialektik der Aufklärung* (Frankfurt/M: Fischer)

Jonas H 1979 *Das Prinzip Verantwortung* (Frankfurt/M: Insel)

Jungk R 1977 *Der Atom-Staat* (München)

Kekes J 1990 *Facing Evil* (Princeton: Princeton University Press)

Kernohan A 1995 Rights against polluters *Environ. Ethics* **17** 245–57

Kermich C 2015 Specifying the concept of future generations for addressing issues related to high-level nuclear waste *Sci. Eng. Ethics* **22** 1797–811

Köhnke D, Reichardt M and Semper F (ed) 2017 *Zwischenlagerung hoch radioaktiver Abfälle* (Basel: Springer Nature)

Lipscomb M and O'Connor R 2002 Democracy and Communicative Rifts: Foucault, Fish, and Yucca Mountain *Admin. Theory Prax.* **24** 393–414

Löfquist L 2019 Is there a universal human right to electricity? *Int. J. Human Rights* **24** 711–23

Luhmann N 1991 *Soziologie des Risikos* (Berlin)

MacIntyre A 1981 *After Virtue* (London: Duckworth)

Mouffe C 2005 *On the Political* (London: Routledge)

Nussbaum M 2018 *The Monarchy of Fear* (Simon and Schuster)

Olson M 1965 *The Logic of Collective Action* (Harvard: Harvard University Press)

Ott K 2003 Reflections on discounting: some philosophical remarks *Int. J. Sustain. Dev.* **6** 7–24

Ott K 2017 Normative pragmatics: approach, promise, outlook *Transcendental Arguments in Moral Theory* ed J Brune, B Stern and M Werner (Berlin: De Gruyter) pp 213–29

Ott 2020 Zur Einlagerung hochradioaktiver Reststoffe aus ethischer und politischer Sicht: Eine Bestandsaufnahme *Competing Knowledges—Wissen im Widerstreit* ed H Anna (Berlin: DeGruyter) pp 171–88

Ott K and Semper F 2017 Nicht von meiner Welt. Zukunftsverantwortung bei der Endlagerung von radioaktiven Reststoffen *GAIA* **26** 100–2

Ott K and Budelmann H 2017 Oder vielleicht doch nicht unter der Erde—Überlegungen zur Rolle der Oberflächenlagerung in einer Entsorgungsstrategie *Zwischenlagerung hoch radioaktiver Abfälle* ed D Köhnke, M Reichardt and F Semper (Basel: Springer Nature) pp 11–26

Ott K and Riemann M 2018 'Volenti non fiat iniuriam'—Freiwilligkeit und Bereitwilligkeit bei der Übernahme von Standortverantwortung *Umwelt, Gerechtigkeit, Freiwilligkeit—insbesondere bei der Realisierung eines Endlagers* ed K Ott and U Smeddinck (Berlin: BWV) pp 41–57

Perrow C B 1984 *Normal Accidents: Living with High-Risk Technologies* (New York: Basic Books)

Pigou A C 1932 *The Economics of Welfare* (London: Macmillan)

Rawls J 1972 *A Theory of Justice* (Harvard: Harvard University Press)

Schulze W, Brookshire D and Sandler T 1981 The social rate of discount for nuclear waste storage: economics or ethics? *Nat. Resour. J.* **21** 811–32

Shrader-Frechette K 1993 *Burying Uncertainty* (Berkeley, CA: University of California Press)

Spaemann R 2011 *Nach uns die Kernschmelze* (Stuttgart: Klett-Cotta)

StandAG (Standortauswahlgesetz) 2017 *Gesetz zur Suche* und *Auswahl eines Standortes für ein Endlager für hochradioaktive Abfälle (in German: Act for the Search for, and the Selection of, a Site for a Repository for Highly Radioactive Wastes)* https://gesetze-im-internet.de/standag_2017/ [accessed 11 April 2021]

Stenius E 1960 *Wittgenstein's ‚Tractatus‘* (Oxford: Blackwell)

Taylor P 1961 *Normative Discourse* (Englewood Cliffs, NJ: Prentice-Hall)

Toulmin S 1958 *The Uses of Argument* (Cambridge: Cambridge University Press)

Themann D, Di Nucci R and Brunnengräber A 2021a Alles falsch gemacht? Machtasymmetrien in der Öffentlichkeitsbeteiligung bei der Standortsuche für ein Endlager *Forschungsjournal Soziale Bewegung Plus* **34** 1–10

Themann D, Schwarz L, Di Nucci R and Brunnengräber A 2021b Power over, power to, und power with bei der Standortsuche für ein Endlager *Forschungsjournal Soziale Bewegung Plus* **34** 1–26

Unnerstall H 1999 *Rechte zukünftiger Generationen* (Würzburg: K&N)

Weston A 2012 *Mobilizing the Green Imagination* (Gabriola Island: New Society Publishers)

Ziegler R 2020 *Innovation, Ethics and Our Common Future* (Cheltenham: Edgar Elgar)

IOP Publishing

Nuclear Waste
Management, disposal and governance
Klaus-Jürgen Röhlig

Chapter 15

Geology, engineering, and society: repository siting as a socio-technical problem

Klaus-Jürgen Röhlig

In several of the failures or setbacks of national nuclear waste management programmes mentioned in part I, chapter 9 'Decision-making in the presence of uncertainty'. Safety Case as a tool supporting the development of deep geologic repository' as well as in part II, chapters 13 'Role of stakeholder involvement in the implementation of radioactive waste management projects' and 16' The German case for dealing with high-level radioactive waste: taking a socio-technical approach to address a socio-technical problem—chances and risks' the choice of locations or sites for nuclear waste management facilities ('siting'), particularly for deep geologic repositories, played a specific role. The history of the German (now abandoned) SNF/HLW disposal project at Gorleben (see part II, chapter 16) is perhaps the most spectacular example of the conflict potential of siting, but there were several more, e.g. the resistance against the early site investigation attempts in the 1980s in Sweden and the conflict around the Wellenberg site in Switzerland (for both, see part II chapter 13). Also, the choice of the Bure site (also called the Meuse/Haute-Marne site) amongst the candidate sites in the departments of Meuse, Haute-Marne, Gard, and Vienne (Andra 2016, footnote 3) was apparently not without controversy (Gras and Massou 2006). Other examples are the back and forth around the Yucca Mountain site (USA, see NWTRB 2021) and the rejection of the project intended to establish a site-specific underground research laboratory, the 'Sellafield Rock Characterisation Facility' (RCF) in the UK (CoRWM 2006, pp 3 f.), as well as the recent criticisms of the siting process initiated in the Czech Republic (SÚRAO 2020, Zachová 2021).

Many approaches to siting taken in the past were focussed on finding a geologic setting that was suitable for repository construction and promising with respect to the long-term (post-closure) safety of the disposal facility to be constructed. Socio-economic considerations related to infrastructure, land-use, and societal impact also played a role. Appendix I of the IAEA's Specific Safety Guide 'Geological Disposal

Facilities for Radioactive Waste' (IAEA 2011) provides a summary of the considerations and criteria relevant to such a process. However, questions of fairness, acceptability, and acceptance are intentionally not addressed in this Guide—and it is probably fair to say that such issues were underestimated in the failed siting attempts mentioned above. This is not to say that these failures could have been avoided if these issues had been better addressed or understood, but it is to say that considering them is a necessary (but not sufficient) condition for such siting processes in today's democratic societies, with their apparently increasing mistrust in traditional decision-making performed by elected representatives or by administrative bodies.

As a consequence, modern siting processes need to account for technical and socio-economic necessities, but, at the same time, strive for acceptability and acceptance. We understand acceptability to mean compliance with ethical and legal standards, while acceptance means that a siting process and its outcomes are accepted or at least tolerated by important actors, such as national and local democratic representatives, civil society organisations, and the populations in the areas affected. Ideally, such actors would consider that an ethically and legally acceptable siting process is fair and appropriate and might therefore at least tolerate its outcomes, even if they are not considered favourable by some actors, but of course it cannot be guaranteed that acceptability and acceptance go together.

Apparently, there are two issues that are important in that respect: one is to design a process which is, and is perceived as, fair, transparent, and comprehensible, see part II, chapter 13 'Role of stakeholder involvement in the implementation of radioactive waste management projects.' The other is to aim for (perhaps formal) consent, but the question arises of what that exactly means. The early and broad engagement of actors (which ones?), the inclusion of veto rights against siting decisions (e.g. for local parliaments or the local populations of affected areas or municipalities), as well as tenders seeking volunteer municipalities are possible approaches.

Obviously, such consent does not necessarily coincide with a favourable geology and favourable socio-economic factors, and it might be challenging to develop approaches combining the two issues. DAEF (2014) distinguishes between siting concepts based on voluntariness and safety-driven approaches which are primarily orientated by geologic features relevant for safety and further recognises: 'In both concepts, an area of tension arises between the primacy of safety on one hand and the achievement of sufficient consent [...] in the affected municipalities or regions on the other.' (translation by the author). How this is addressed will also depend on several other factors, such as the geological setting of a country, its population density and distribution, as well as its political culture and historical background. Below, we discuss selected examples of siting processes in order to illustrate important features.

Sweden and **Finland** are located on the Baltic or Fennoscandian Shield, which is characterised by crystalline bedrock. Early studies carried out in Sweden revealed that a considerable part of the country's geology had the potential to host a repository (SKB 2011, figure 4-2). This provided relatively large degrees of freedom for a siting process based on voluntarism, which was the approach taken following confrontations with the local population in connection with early site investigation attempts. Feasibility studies were carried out in several municipalities, which volunteered in response to requests sent out to every municipality in Sweden. In the ensuing process, two municipalities (Oskarshamn with the Laxemar site and Östhammar with the Forsmark site) remained as candidates for a repository site. The others either withdrew their applications or were ruled out due to scientific or feasibility reasons (SKB 2011, 2021). The Swedish implementer SKB finally compared the two sites based on the following principles:

'The strategy established for the selection process was based on the purpose of the final repository project and formulated in two points:

1. The site that offers the best prospects for achieving long-term safety in practice will be selected.
2. If no decisive difference is found between the sites in terms of their prospects for achieving longterm safety, the site that is judged to be the most favourable from other aspects for accomplishing the final repository project will be selected.' (SKB 2011)

The factors considered for comparison were:

- Safety-related site characteristics,
- The technology used for execution,
- Societal resources, and
- Health and environmental issues (SKB 2011).

At Forsmark, there are lower groundwater flow rates and a lower rate of occurrence of water-conducting fractures than at Laxemar. As a consequence, Forsmark ranked higher in terms of safety and technology for execution, although the higher rock stresses at Forsmark are also causing risks with respect to technical execution. Mainly due to the higher thermal conductivity, a repository at Forsmark can be smaller than at Laxemar, which leads to lower costs. On the other hand, several so-called 'external factors,' in particular, the vicinity to the interim storage facility Clab, favored Laxemar. No decisive differences between the two were identified with respect to environmental, health, and societal issues.

'In an overall evaluation, SKB concludes that a final repository at Forsmark offers significantly better prospects for achieving long-term safety than a final repository Laxemar. In accordance with the strategy for site selection, SKB has therefore decided to locate the final repository at Forsmark' (SKB 2011).

It should be noted that both municipalities remained part of the process on a voluntary basis until the final decision. Another important point is that both municipalities already host nuclear facilities (nuclear power plants in both municipalities, the LILW repository at Forsmark, and the SFN interim storage Clab in the Oskarshamn municipality, where the encapsulation plant for SNF conditioning will also be built) which might be an important factor affecting local acceptance.

In summary, in the opinion of the author it is fair to say that both voluntariness and safety were necessary conditions for siting Sweden's SNF repository in a process in which the geologic conditions offered enough flexibility to focus on municipalities with high acceptance for the project, including those already hosting nuclear facilities. Political culture and SKB's continuing efforts to engage the local population (see e.g. Soderblom 2014) were also of importance.

Similar observations can be made about the situation in Finland, although the approach to siting which resulted in the choice of the Olkiluoto site for SNF disposal was different: the starting point was the geoscientific criteria, and there was no tender for volunteer municipalities (McEwen 2000, Kojo 2009). However, acceptance by the Eurojoki municipality in which Olkiluoto is situated was crucial.

Canada (see the case study in part II, chapter 13 'Role of stakeholder involvement in the implementation of radioactive waste management projects') is taking an approach to siting which shows various similarities to that of Sweden: it is to be expected that Canada's geology will offer ample possibilities to host a repository (NWMO 2010). After an information phase, communities were encouraged to express their interest in more detailed information about the possibility of hosting a repository, including an initial screening to address issues such as land availability, potential usage conflicts, and geologic conditions obviously disadvantageous with respect to safety. If the communities passed the screening, they were given more detailed information. If they remained interested, a 'Preliminary Assessment' of their suitability was carried out. The assessment addressed safety, potential benefits for the community and its neighbours, and perspectives concerning acceptance. Recently, the investigations were narrowed down to two out of the more than 20 communities which initially expressed their interest. Although others might also have the potential for hosting a repository, the implementer NWMO (the Nuclear Waste Management Organization) obviously opted for those communities with the highest potential. This is evidenced in the documentation by statements such as:

'The geoscientific desktop preliminary assessment showed that the geological setting in the area of [...] has a number of favourable geoscientific characteristics for hosting a deep geological repository for used nuclear fuel. However, the assessment revealed that there are areas that have more potential than others to satisfy NWMO's geoscientific site evaluation factors.' (NWMO 2014)

For 2023, the implementer NWMO plans to 'identify a single preferred site and move into the licensing and regulatory process' (NWMO 2021).

Besides of the similarities with the Swedish process (promising geology over large areas, a process based on voluntariness and safety considerations, and a request that communities actively apply to become part of the process), there are also differences: The Canadian Adaptive Phased Management approach (NWMO 2005, see also part I, chapter 7 'Disposal and other conceivable strategy 'endpoints' for different types of waste') offers more flexibility e.g. with respect to the implementation of interim storage. Also, the two communities remaining in the process offer different geologic situations: crystalline rock at Ignace in northern Ontario and argillaceous limestone at South Bruce in southern Ontario. As a consequence, the concepts for the potential repositories will be different from each other, which might (but not necessarily will) complicate the identification of a preferred site.

In the **UK**, which offers several potential host rocks, consent at the community level is also a necessary condition for hosting a repository. In parallel with consultations with communities, a 'National Geological Screening' was carried out, the results of which helped communities to assess their geological situation (RWM 2020).

In contrast to the countries mentioned above, **Switzerland** and **Germany** are following a stepwise approach which is focussed on safety (and thus on geology) and, secondarily, on spatial planning. Participation is an essential part of the procedures, but voluntariness or the consent of local stakeholders is not foreseen. While Switzerland applies a legal instrument that was established for spatial planning (Sectoral Plan, in German 'Sachplan,' BFE 2011, see the case study in part II, chapter 13 'Role of stakeholder involvement in the implementation of radioactive waste management projects'), Germany passed a 'Site Selection Act' specifically tailored for the task at hand (StandAG 2017, see also part II, chapter 16 'The German case for dealing with high-level radioactive waste: taking a socio-technical approach to address a socio-technical problem—chances and risks'). In both cases, decisions are to be made by parliament: in Switzerland, the siting decision is connected with a licensing decision ('Rahmenbewilligung') of the government ('Bundesrat') which has to pass in parliament ('Bundesversammlung') and might then become subject to an optional referendum ('fakultatives Referendum'), i.e. a referendum which will only take place if there is an official demand by a popular initiative. In Germany, there are three siting stages, during which the range of areas considered will be narrowed down and the degree of site investigation detail will increase. Each of them requires a parliamentary decision.

Another notable difference is that the German Site Selection Act explicitly requires the investigation of three host rock types (rock salt, claystone, and crystalline rock), all of which are abundant in the country. Moreover, the Act requires the identification of the 'best possible site' in terms of safety. At first glance, this poses an epistemic problem, but the Act further specifies that this is going to be the site which is identified in the process defined by the Act. This process is based on predefined geoscientific criteria as well as on safety assessments, but of course also involves spatial planning. Nevertheless, in the case in which candidate sites with different host rocks (and thus requiring different repository concepts) remain at the end of the process, their comparison and the identification of the 'best' site might become a methodological problem. In Switzerland, the process has already revealed that a site with a clay host rock will be chosen for SNF/HLW disposal, although the 'Sachplan' did not prescribe the choice of a host rock. Questions of LILW disposal, which also play a role in both siting processes, are not addressed here.

In the **Czech Republic**, SNF disposal in crystalline rock is foreseen. Siting is being performed based on geoscientific and spatial planning criteria and considerations and involves the participation of local stakeholders. Important decisions are to be taken by the government (SÚRAO 2020).

In the opinion of the author, it is not advisable to derive general conclusions on the quality or success chances of these or other siting process for HLW/SNF disposal. This is due to the high variation of geologic and societal conditions in the countries discussed and the fact that only a limited number of countries are actively working on such processes (see part I, chapter 8 'Geologic ('deep') disposal of high-level and other long-lived waste: host rocks, concepts, current international status'). However, the following observations can be made:

- Present-day siting processes account for safety-relevant geoscientific as well as for spatial planning issues and strive for fairness by involving interested and affected actors.
- They vary concerning the weight attached to these issues.
    - In particular, there are differences about whether, and to which degree, voluntariness or the consent of affected local actors play a role.
    - There is also variation about where to start from (voluntariness or geoscientific and planning considerations) and how the interplay between the two latter considerations forms the process.
- Further variation exists concerning the instruments utilised in the process. Apparently, the countries working without voluntariness or consent concepts attempt to legitimise the process by means of participation formats and by decision processes involving the legislature (Switzerland, Germany) or the executive (Czech Republic). Countries with 'consent-type' approaches rely on the interaction between implementing organisations and local stakeholders.
- Factors that influence the above might be the geologic setting of a country as well as the general attitude of the population to large infrastructure projects and to nuclear installations (not discussed in this chapter). In addition, the country-specific political setting, history, culture, and legal system play important roles.

Bibliography

Andra (Agence nationale pour la gestion des déchets radioactifs) 2016 *Safety Options Report—Operating Part (DOS-Expl)* CG-TE-D-NTE-AMOA-SR1-0000-15–0060 https://international.andra.fr/sites/international/files/2019-03/Safety%20Options%20Report%20-%20Operations_2.pdf

BFE (Bundesamt für Energie) 2011 *Sachplan geologische Tiefenlager. Konzeptteil* 2 April 2008 *(Revision vom 30 November 2011)* (in German, Sectoral Plan for Deep Geological Repositories. Conceptual Part)

CoRWM (Committee on Radioactive Waste Management) 2006 *Managing our Radioactive Waste Safely. CoRWM's Recommendations to Government* https://gov.uk/government/uploads/system/uploads/attachment_data/file/294118/700_-_CoRWM_July_2006_Recommendations_to_Government_pdf.pdf

DAEF (Deutsche Arbeitsgemeinschaft Endlagerforschung, German Association for Repository Research) 2014 *Aspekte eines Standortauswahlverfahrens für ein Endlager für Wärme entwickelnde Abfälle* (in German, Aspects of a Site Selection Procedure for a Repository for Heat-generating wastes in Germany) http://daef2014.org/DAEF/assets/daef_broschuere_okt_2014.pdf

Gras A and Massou F 2006 La gestion des déchets nucléaires *La Gazette de la société et des techniques* **36** p 3 http://www.annales.org/gazette/Gazette_web_36bis.pdf [accessed 22 May 2021]

IAEA (International Atomic Energy Agency) 2011 *Geological Disposal Facilities for Radioactive Waste* Specific Safety Guide, IAEA Safety Standards Series No. SSG-14

Kojo M 2009 The strategy of site selection for the spent nuclear fuel repository in Finland *The Renewal of Nuclear Power in Finland (Energy, Climate and the Environment Series)* ed M Kojo and T Litmanen (London: Palgrave Macmillan)

McEwen T 2000 *The Site Selection Process for a Spent Fuel Repository in Finland—Summary Report* POSIVA 2000-15 https://inis.iaea.org/collection/NCLCollectionStore/_Public/32/030/32030717.pdf [accessed 24 May 2021]

NWMO (Nuclear Waste Management Organization) 2010 *Moving Forward Together: Process for Selecting a Site for Canada's Deep Geological Repository for Used Nuclear Fuel* https://nwmo.ca/~/media/Site/Files/PDFs/2015/11/04/17/34/1545_processforselectingasiteforcan.ashx?la=en

NWMO (Nuclear Waste Management Organization) 2014 *Phase 1 Preliminary Assessments Summary Findings and Decisions* https://nwmo.ca/~/media/Site/Files/PDFs/2015/11/04/17/37/2445_phase_1_preliminary_assessments_-_summary_findings.ashx?la=en

NWMO (Nuclear Waste Management Organization) 2021 *Implementing Adaptive Phased Management 2021 to 2025* https://nwmo.ca/~/media/Site/Reports/2021/03/15/21/05/Implementing-Adaptive-Phased-Management-2021-to-2025.ashx?la=en

NWTRB (U.S. Nuclear Waste Technical Review Board) 2021 *Six Overarching Recommendations for How to Move the Nation's Nuclear Waste Management Program Forward. A Report to the U.S. Congress and the Secretary of Energy* https://nwtrb.gov/docs/default-source/reports/six-recommendations-report_508c.pdf?sfvrsn=18

RWM (Radioactive Waste Management) 2020 *Guidance. Geological Disposal. A Permanent Solution for the UK's Higher-Activity Radioactive Waste* https://gov.uk/guidance/geological-disposal

SKB (Svensk Kärnbränslehantering AB, Swedish Nuclear Fuel and Waste Management Company) 2011 *Site Selection—Siting of the Final Repository for Spent Nuclear Fuel* SKB Report R-11-07 www.skb.com/publication/2204001/R-11-07.pdf

SKB (Svensk Kärnbränslehantering AB, Swedish Nuclear Fuel and Waste Management Company) 2021 *How Forsmark Was Selected* www.skb.com/future-projects/the-spent-fuel-repository/how-forsmark-was-selected/

Soderblom A 2014 *How Swedish Communities Organised Themselves in Reviewing a Safety Case* in: *The Safety Case for Deep Geological Disposal of Radioactive Waste: 2013 State of the Art. Symp. Proc., 7–9 October 2013, Paris, France* OECD, NEA/RWM/R(2013)9

StandAG (Standortauswahlgesetz) 2017 *Gesetz zur Suche und Auswahl eines Standortes für ein Endlager für hochradioaktive Abfälle (in German, Act for the Search for, and the selection of, a site for a repository for highly radioactive wastes)* www.gesetze-im-internet.de/standag_2017/

SÚRAO (Správa úložišť radioaktivních odpadů, Radioactive Waste Repository Agency) 2020 *Site hortlisting. Deep Geological Repository Site Shortlisting Information Page* https://surao.cz/en/site-shortlisting/

Zachová A 2021 *Czech Municipalities Fight Against Nuclear Waste Repository EURACTIV* https://euractiv.com/section/politics/short_news/czech-municipalities-fight-against-nuclear-waste-repository/

Chapter 16

The German case for dealing with high-level radioactive waste: taking a socio-technical approach to address a socio-technical problem—chances and risks

Armin Grunwald

16.1 The burden of history

Technology conflicts are not a new phenomenon in modern times, as the well-known example of the Luddites at the beginning of the 19th century demonstrates. They destroyed textile production machines because of (justified, as history showed) concerns that these would make their jobs superfluous. However, since the 1960s, severe technology conflicts and deficits of legitimacy related to new technologies have reached a more dramatic level in many Western countries. The best-known example, that of the many conflicts around nuclear power (e.g. Sovacool 2011), leads directly to the issue of (in particular, high-level) radioactive waste disposal. In many Western countries, strong and partially ongoing conflicts arose with the appearance of this issue (Brunnengräber *et al* 2019). In Germany, the siting process related to the salt dome at Gorleben was among the worst conflicts in recent decades. In spite of the complete relaunch of the search for a disposal site after Germany's decision to phase out nuclear power following the Fukushima disaster (section 16.2), this conflict still burdens the current site selection process.

The origin of this conflict is grounded in the history of nuclear power in Germany (Radkau/Hahn 2013). While nuclear power had a positive resonance in the German elites, media, and population in the late 1950s and in the 1960s and was regarded an essential element of a good future, things started changing in the 1970s, following the 1968 students' movement. Together with the emerging green and peace movements of those years, a strong antinuclear movement established itself and formed the origins of today's environmental parties, not only in Germany but also beyond. Initially, this was characterized as a grass-roots civil society and extra-parliamentary

doi:10.1088/978-0-7503-3095-4ch16

movement, which did not accept the decisions made by governments and federal authorities using the representative model of democracy. Demonstrations against planned new nuclear power plants, such as those that took place in Wyhl close to Freiburg in Southern Germany, exerted pressure on decision-makers. Among the first successes of the antinuclear movement was preventing the construction of the Wyhl power plant in 1976. The Chernobyl accident in 1986 fueled resistance and contributed to a broadening of the antinuclear movement's support to a larger portion of society.

The antinuclear movement raised several arguments against nuclear power. First, the movement raised the issue of possibly deficient safety and the fears of accidents and disasters. Second, concerns related to cancer and other health issues due to ionizing radiation close to nuclear power plants and around nuclear transports were raised. Third, many suspicions against governments and authorities in relation to big companies with respect to the possible military use of nuclear technologies for atomic weapons played a major role, in particular with regard to German history in two world wars. And fourth, the *nonexistence* of a plan for dealing with the (in particular, high-level) radioactive waste began to become a major issue.

In the early stages of exploring and establishing nuclear power as a basis for the German electricity supply, experts and decision-makers were not particularly concerned about waste disposal. In the 1960s, the first ideas concentrated on disposal sites in salt domes in Lower Saxony, initiating a series of scientific investigations into salt domes as promising places for the final disposal of high-level waste (HLW). Based on a scientific analysis, four sites were identified as good candidates. However, in 1977 the Prime Minister of Lower Saxony, Ernst Albrecht, made a decision for a site beyond those four, which then was approved by the regional parliament. The final disposal site was to be established at Gorleben close to the border with the German Democratic Republic, together with an industrial plant for interim storage, the recycling (reprocessing) of spent nuclear fuel (SNF) for reuse in nuclear power plants of the fast breeder type, and waste conditioning for disposal. The combination of distrust of politics and authorities, the protest against nuclear power, and the experience of a decision seemingly based neither on scientific evidence nor on transparent political deliberation formed an explosive mixture, which led to one of the worst conflicts in West German history since World War II.

From 1995 to 2011, several transports of Castor containers loaded with SNF from nuclear power plants, but mostly with HLW from reprocessing in La Hague (France) were brought to the interim storage site at Gorleben to be finally disposed of in the neighboring salt dome after the disposal mine was finished. In parallel, the mine to be used for the final disposal in the salt dome was explored and prepared. The protests against these transports regularly led to riots and violence. Thousands of policemen were needed to safeguard each of the transports, while tens of thousands of demonstrators from all over Western Germany created and exerted various forms of protest and resistance. Pictures from these civil-war-like events spread around the world. In 2000, after the red/green coalition in Germany's federal government was established, the exploration of the salt dome was interrupted and only resumed for a short time period (2010–2012). Effectively, the issue of dealing with the radioactive HLW became 'frozen' in political deadlock and unsolved conflict until 2013.

More than 40 years after the decision by Ernst Albrecht, it is more than clear that the first attempt to determine a safe repository for HLW in Germany was a failure. The selection process and the decision in favor of Gorleben were neither transparent nor science-based, but motivated resistance far beyond the region concerned. During this conflict, trust in democracy and its institutions was damaged beyond only small parts of the population. This loss of trust in the democratic state is the most significant heritage of this first approach to dealing with high-level radioactive waste—and has raised many issues for learning. History remains a heavy burden in the ongoing process.

16.2 A window of opportunity for a relaunch

The nonexistence of a solution for dealing safely with HLW was a major argument of the antinuclear movement. Hence, the willingness of the movement to agree to a relaunch of the siting process after the devastating Gorleben story was limited. Its concern was that a solution could weaken its own position and, instead, give a positive impetus to the proponents of nuclear power. This argumentative deadlock disappeared after the Fukushima disaster in early 2011. The German Government quickly decided to phase out nuclear power completely until 2022, supported by a wide consensus among the population. Simultaneously, the transformation of the energy system to renewable energies, called *Energiewende*, which was envisaged anyway, was to be accelerated.

This decision allowed large parts of the antinuclear movement to rethink their positions on the waste issue without having to be afraid that a constructive attitude could contribute to a prolongation of the use of nuclear power in Germany. In particular, the Green Party (*Bündnis* 90/*DIE GRÜNEN*), deeply rooted in the history and positions of the antinuclear movement, declared its willingness to support a relaunch of the HL waste disposal site selection process. In this way, a broad parliamentary consensus that included the Christian Democrats and the Christian Social Union (CDU/CSU), the Social Democrats (SPD), the Liberal Democrats (FDP), and the Green Party became possible. Only the left-wing party DIE LINKE and parts of the antinuclear movement remained outsiders, for different reasons. One of the reasons was, and to some extent still is, persistent mistrust which suggested that the relaunch would not be taken seriously and could, instead, lead back to a revival of the Gorleben site.

Broad parliamentary consensus in the German *Bundestag* and in the state representation (*Bundesrat*) made it possible to approve the 'Act on the Search for and Selection of a Site for a Disposal Facility for High-Level Radioactive Waste,' in short, the Site Selection Act (in the following, the German abbreviation StandAG will be used). One of the major regulations was the installation of the 'German Commission on the Storage of High-Level Radioactive Waste' (the *Endlagerkommission*, in the following, referred to as the 'Commission')[1]. It was

[1] Using the term 'storage' instead of 'disposal' should imply an openness with regard to the management strategy, for which several options were investigated by the Commission.

mandated to recommend a management strategy for HLW and to develop rules, criteria, and procedures for a completely new siting process to search for the site with the best possible safety in Germany (Liebscher *et al* 2020). To achieve this, the Commission was obliged to take into account lessons from the past and failed approaches to solving the HLW problem (section 16.1). The StandAG in its initial form (from 2013) and ethical considerations motivated the Commission to determine the following requirements for the path towards safe final disposal (Commission 2016a, 6):

- The search for the disposal path, final repository site, and concept must be primarily directed towards the goal of finding the safest form of disposal of high-level radioactive waste, in particular, given what we know today: safety takes precedence.
- The radioactive waste produced domestically must be placed in safe final disposal in Germany.
- The form of disposal must be designed in such a way that it does not place a permanent burden on future generations but, rather, provides for a safe final condition for the disposal of all high-level radioactive waste.
- The solution must allow for the option of a decision that deviates from the option recommended here. It must be possible to correct mistakes. Unnecessary irreversibilities must be avoided. Before irreversible decisions, or decisions that can only be revised at great expense and effort, are taken, a transparent and scientifically validated evaluation must be carried out with the participation of the public and the relevant committees.
- The whole process must be transparent, with the significant participation of the public and the regions affected, and must be designed as a self-reflecting system.

The Commission was composed of 32 members plus two chairpersons. Sixteen members were appointed from among politicians, half as members of the *Bundestag*, half nominated by the states via the *Bundesrat*. Eight members from the scientific community mainly represented geoscience and geotechnology as well as legal science, complemented by one ethicist from the field of technology assessment. The other eight members were nominated by stakeholder groups: industry, labour unions, churches, and environmental initiatives. The siting process to be developed by the Commission should allow determination of the site with the *best possible safety in Germany* without any other predetermination. This was laid down in the amendment of the StandAG, which was approved by the Bundestag and the Bundesrat based on the Commission's recommendations in 2017 (StandAG 2017 in German and English versions), and which guides the current siting process:

> The purpose of the site selection procedure is to identify in a participatory, science-based, transparent, self-questioning and learning process a site for the disposal of high-level radioactive waste generated in Germany with the best possible safety for a disposal facility pursuant to § 9a(3) sentence one of the

> Atomic Energy Act in the Federal Republic of Germany. The site with the best possible safety is the site identified on the basis of a comparative procedure from among the suitable sites in the respective phase in accordance with the corresponding requirements of this Act ensuring the best possible safety for the long-term protection of man and the environment against ionising radiation and other harmful effects of such waste for a period of one million years. This shall also include avoiding undue burdens and obligations for future generations (StandAG 2017, §2).

Following a review of the other options available for the final disposal of HLW, the commission recommended disposing of the waste in a deep geological formation while observing ethical principles regarding future generations (e.g. Hocke 2015) and designing the site selection as a participatory and learning process. This obligation leads to the requirements to take care of the reversibility of decisions and the possibility of correcting mistakes and to establish monitoring (Hocke *et al* 2012) at different levels. Learning will also help to build trust in the procedure, which faces the persistent mistrust of parts of society. Therefore, the new site selection process must be built on transparency and public participation, according to the Commission (see also Grunwald 2015).

The very idea of this open-ended procedure for selecting the final repository site is that the site with the best possible safety can be established during the procedure only, because this result involves scientific, technological, and ethical issues in an inherently interrelated socio-technical manner. This procedural approach is the logical consequence of the conflict-ridden final disposal efforts seen in Germany in the past, because it prevents a premature commitment to a site *before* its exploration has been completed (Kanitz 2020). The siting process is an open-ended procedure, which prevents any suspicion that it it merely being used to confirm prior, politically motivated decisions for a site.

16.3 The siting process

The recommendations of the Commission (summary in English Commission 2016a; full text Commission 2016b) were widely adopted by the *Bundestag* and the *Bundesrat*, and included in the modified version (novel) of the StandAG, which prescribes the objective of the siting process:

> At the site to be selected, disposal shall take place in deep geological formations in a disposal mine built for this purpose with the aim of final sealing. Retrievability during the operating phase of the disposal facility and recoverability for 500 years after the planned sealing of the disposal facility shall be provided for (StandAG 2017, §1,4).

Shortly after the new version of the StandAG was approved in 2017, the new site selection process was started with a 'white map' of Germany. The entire area of Germany was considered for potentially suitable host rock formations and potentially suitable regions. Any predetermination in favour of or against a specific region was to be strictly avoided. The geological safety criteria, recommended by the Commission, were made part of the StandAG by the *Bundestag*. They have to be applied to all the rock types that are theoretically suitable for the hosting of a final repository: 'In Germany, rock salt, claystone and crystalline rock shall generally be considered as host rocks for the disposal of high-level radioactive waste' (StandAG 2017; §1,3).

16.3.1 The institutional configuration of the selection process

The siting process needs institutions to perform various tasks: applying the criteria and drawing conclusions, informing the public, organizing public participation, supervising the activities for compliance with the StandAG and other regulations, taking care of the high ambitions with regard to learning, and so forth. Three institutions are involved at the core of the siting process, while political bodies such as the *Bundestag*, state governments represented in the *Bundesrat*, and the German ministry for the Environment, Nature Conservation, and Nuclear Safety (BMU) take care of the political and regulative frames[2]:

- The *Bundesgesellschaft für Endlagerung* (BGE) is mandated by the federal government to perform tasks in the final disposal of radioactive waste. The federally owned company was founded for this purpose in July 2016. BGE is looking for a site for a repository for high-level radioactive waste. Its tasks include the preparation of proposals for a site for high-level radioactive waste in accordance with StandAG, the preparation of the respective preliminary safety investigations, and the construction and operation of the repository after site selection (from the BGE homepage; see https://www.bge.de/en/ [accessed 8 April 2021]; see also Kanitz 2020).
- The *Bundesamt für die Sicherheit der nuklearen Entsorgung* (BASE, Federal Office for the Safety of Nuclear Waste Management) controls the search for a repository site on behalf of the federal government and organizes public participation. The BASE will assess all the proposals submitted by the BGE in the site selection process, organize the participation of the public, and forward its results to the BMU. At the end of the site selection process, the BASE will make a proposal for a repository site (from the BASE homepage; see https://www.base.bund.de/EN/ [accessed 8 April 2021]).
- The *Nationale Begleitgremium* (NBG, the National Citizens' Oversight Committee) is an institutional innovation without previous models (Smeddinck 2020). Its tasks reflect the specific situation of the site selection in Germany, in particular, the burden of history (section 16.1) and the need to build trust. Because of its specific nature, the NBG will be described in more depth below (section 16.3.4).

[2] Only the responsibilities of the organisations which relate to HLW disposal are mentioned here.

According to the lessons learned from the problematic German history with previous approaches to dealing with radioactive waste, public participation and transparency are prescribed for all steps of the entire selection process:

> Under this Act, the Federal Office for the Safety of Nuclear Waste Management shall ensure that the public is informed duly and throughout the duration of the site selection procedure of the objectives of the project, the means and progress of its implementation as well as its likely impacts and of forms of participation provided for. This should be done in a dialogue-oriented process (StandAG 2017, §3,2).

Public participation measures prescribed by the StandAG are to be organized by BASE. NBG observes and supervises these activities but also develops own opportunities for dialogue with various decision-makers, stakeholders, and citizens.

16.3.2 Stages, phases, and steps of the process

According to the recommendations of the Commission (2016a), the path towards finally disposing of HLW shall be divided into stages, seen from today's perspective. Within each of these stages, continuous monitoring (Hocke *et al* 2012) shall take place to allow learning and also, as a possible consequence, the modification of the respective next steps in the case of new insights. The regulations laid down in the StandAG (2017) focus in depth on stage 1, the site selection procedure.

Stage 1–Site selection procedure: Starting by considering the whole territory of Germany based on currently available data, the sites for subsequent surface exploration have to be identified by the BGE in the first phase of stage 1. Regions are to be excluded where, based on the geological criteria (section 16.3.3), final disposal does not appear feasible from the outset. The BGE draft interim report on potentially promising subareas (*Teilgebiete*), as the first step in phase 1, has to be published (which happened in September 2020) and made an object of public scrutiny and feedback (see section 16.3.4). The BGE has to take this feedback into account in its final report.

Representative preliminary safety analyses (see section 16.3.3) form the second step in phase 1. They will be compiled by the BGE to narrow down the potential subareas received from step one to potential sites to be considered in depth in the further process. Again, this step will be based on available data only. Based on these analyses, the BGE will propose regions for surface exploration (*Standortregionen*). BASE will examine the BGE proposal and make a proposal to BMU, which will submit the proposal to the *Bundestag* in the form of a draft bill. The approval of this bill concludes phase 1.

In phase 2, surface exploration (StandAG 2017, §15) will be carried out in the selected potential regions. The data gained will be used for a more detailed application of the geological safety criteria. BGE will analyse the results, and on this basis, further develop and deepen the preliminary safety analyses (see section 16.3.3) from the preceding phase. The population in the affected regions will be involved on a regular basis through regional conferences accompanied by other formats. The result of this phase will be a list of a few candidate locations for receipt of the final repository, which again has to be approved by BASE, BMU, and finally by the *Bundestag*.

In phase 3, the BGE performs underground exploration (StandAG 2017, §17) of the remaining potential sites. It subsequently submits a report on the findings of the exploration and its conclusions to the BASE. Transparent documentation and reasons for all the steps taken and appraisals made have to be provided and published. The concluding comparative analysis of the sites, based on the evaluation of the data gained by underground exploration, and the elaboration of a site proposal, have to involve the public. Finally, a site will be proposed by BASE on which the *Bundestag* will make the decision (StandAG 2017, §20).

Hence, the selection of the site shall take place in a series of steps that gradually narrow down the potential regions and sites based on a continuously improved set of data until the site that offers the best possible safety has been determined (Commission 2016a). While the year 2031 was laid down in the StandAG as the target for making the final decision, several voices, e.g. from the Commission, have expressed doubts about the feasibility of this schedule. Any delay, however, will have consequences for the existing interim storage of HLW. If the selection of the final disposal site takes an unexpectedly long time or a decision is made to change to a different disposal path, it may become necessary to initiate technically, economically, and institutionally complex processes to support prolonged safe interim storage, cf. part I, chapter 6 'Storage and transport'.

Stage 2–Developing and building the disposal mine: The mining development of the site for the disposal of HL radioactive waste comprises planning, approval, and construction processes. It includes provision of proof of the long-term safety of the mine considering the entire disposal system, also including geological barriers and technical concepts. The underground repository has to be built completely with all the required surface and underground technical facilities. The transport logistics needed to deliver the waste to the mine also have to be prepared.

Stage 3–Emplacement of HLW: Optimistically speaking, the emplacement of the waste can begin around the year 2050. The details will depend on the selected final disposal concept and the host rock. At this stage, it has to remain possible to reopen already closed units and to retrieve the waste stored in case of need. This requirement also implies that the containers need to be designed and constructed in a manner that allows possible retrieval.

Stage 4–Observation of the repository mine: For the purpose of monitoring geological developments in the mine, e.g. aspects such as temperature, the stability of the geological formation, and gas formation, the mine shall be observed for a further unspecified timespan after the emplacement of the HL waste. During this stage, the mine remains fully functional and accessible. In case of necessity, the waste containers could still be retrieved. The sealing of the repository mine marks the end of this stage. The goal of safe and maintenance-free containment of radioactive waste in the mine is then reached.

Stage 5–Sealed repository mine: The sealed repository mine can only be further observed from the outside. If required, e.g. in the case of unforeseen developments, the waste containers could be recovered by opening up a new mine adjacent to the initial one. Recovery remains possible as long as the containers are in a recoverable state, given that the documentation is still available and legible.

The entire process will take more than a century, even in the best case. The risk of delay will accompany all stages and could further increase the timespan. While careful data investigation and evaluation as well as participation and reflected deliberation need time and tend to expand the overall time needed, risks to the procedure also increase with time. Even the economic or political stability of society could come under pressure, which would threaten completion of the procedure in good order. Therefore, dealing with this trade-off needs careful consideration.

16.3.3 Safety criteria

The main part of the navigation during the process of reducing the 'white map' of Germany to the one remaining site offering the best possible safety is driven by geological safety criteria, recommended by the Commission (2016a), and laid down in the StandAG (2017). These criteria are exclusion criteria, minimum requirements, and assessment (evaluation) criteria, together with safety requirements and safety analyses. These crucial elements of the process will briefly be described in this section, based on the StandAG. The categorically different class of scientific planning criteria (StandAG 2017, §25), which regulates the impacts of settlements or environmental protection areas, will only be applied in the later stages. Because these must not overrule the geological safety criteria, they will not be described in this paper.

It seems remarkable that the geological criteria are described, and therefore also fixed in the StandAG in depth and in detail. Several examples will be given below; they even mention geochemical or geophysical parameters and abbreviations and prescribe exact thresholds for some of them. These definitions guarantee that the selection process will be science-based and free from political intervention. Hence, they are another reaction to the historical burdens weighing on this issue in Germany (section 16.1).

The set of criteria shall remain the same over the entire selection process. While this regulation is understandable with respect to consistency of approach over time and the comparability of all of the various assessments to be performed, it could cause conflict with the requirements of the learning process. Considering the long timespan between the approval of the StandAG and the determination of the disposal site, it seems not improbable that new scientific insights could give rise to some changes or amendments to the initial set of criteria. In this case, a critical evaluation and careful deliberation will be needed in order to develop the procedure further without causing damage to the selection process.

Exclusion criteria provide reasons why a specific area should, for geological reasons, no longer be considered a candidate for the disposal site, regardless of its properties in other respects. Some examples are (StandAG 2017, §22,2):

large-scale vertical movements: a large-scale geogenic uplift exceeding more than 1 mm on average per year is to be expected during the reference period of one million years;

seismic activity: the local seismic hazard is greater than in seismic zone one according to DIN EN 1998–1/NA 2011–01;

volcanic activity: there is quaternary volcanism or future volcanic activity is to be expected;

age of groundwater: in the rock zones eligible to be used as the containment-providing rock zone or the emplacement zone, young groundwaters have been detected.

Areas that are not touched by the exclusion criteria have to be evaluated with respect to the *minimum requirements* an area must fulfill in order to be investigated further. Areas that do not fulfil these requirements will be excluded from further consideration. Two examples are (StandAG 2017, §23,5):

1. rock permeability: the rock permeability k_f must be less than 10^{-10} m s^{-1} in a containment-providing rock zone; where direct evidence cannot be provided yet in the statements of grounds for the proposals pursuant to §§ 14 and 16, it must be demonstrated that the containment-providing rock zone consists of rock types to which a rock permeability lower than 10^{-10} m s^{-1} can be assigned; the fulfilment of this criterion may also be demonstrated by strata overlying the emplacement zone.

2. thickness of the containment-providing rock zone: the rock zone which is to host the containment-providing rock zone must be at least 100 metres thick; in the case of rock bodies of crystalline host rock of smaller thickness, the evidence of safe containment for the rock area in question may also be provided by the interaction of the host rock with geotechnical and technical barriers; subdivision into several such rock zones within a disposal system shall be permissible.

By applying *geoscientific evaluation criteria*, an assessment can be made of whether there is a favorable overall geological situation in a certain area. The presence of a favorable overall geological situation can be confirmed after a safety-oriented weighing and balancing process that considers the results for all criteria. These have been divided into three classes (StandAG 2017, §24):

(1) The achievable quality of containment and the robustness of the evidence to be expected shall be assessed on the basis of the criteria for transportation

through groundwater, for configuration of the rock bodies, for spatial characterisability and for predictability. These criteria are specified in Appendices one to four.

(2) The preservation of the isolation performance shall be assessed on the basis of the criteria for rock-mechanical preconditions and for the low tendency to form fluid flow paths. These criteria are specified in Appendices five and six.

(3) Other safety-relevant properties shall be assessed on the basis of the criteria for gas formation, for temperature tolerance, for the retention capacity of the rocks of the containment-providing rock zone for radionuclides, for hydro-chemical conditions and for the overburden. These criteria are specified in Appendices seven to eleven.

These rough descriptions and also the more detailed explanations in the appendices already include vague notions such as quality, robustness, low tendency, sufficient, relevant, etc. These notions have to be filled with specific meaning in order to prevent biased interpretation and to allow clear orientation. To this end, the lengthy appendices provide a vast amount of detail to better explain the assessment criteria and to avoid biased interpretation. This exact prescription at the level of geochemical, geophysical, and hydrogeological definitions seems to be a remarkable part of the German case. It mirrors the intention to implement a science-based navigation of the selection process while excluding space for political interpretation or intervention.

Safety requirements shall guarantee that the long-term and safe containment of the radioactive waste can be expected at a site with the foreseen disposal concept. *Safety analyses* are the science- and data-based assessments of the extent to which safe containment of the radioactive waste can be expected by exploiting the geological conditions at the site. While the geological criteria mentioned above address the geological properties of the host rock and its environment, safety analyses have to consider the combination of geological and technical issues. In phase one of stage one of the process, only preliminary safety analyses can be conducted because of the lack of in-depth data (StandAG 2017, §27). In stages two and three, the safety analyses have to be conducted in an increasingly detailed and context-specific manner. Examples of the safety requirements are (StandAG 2017, §26):

1. The radioactive and other pollutants in the waste shall be concentrated and contained … with the aim to keep them away from the biosphere. For a period of one million years, it must be ensured with regard to the protection of human beings and, as far as the long-term protection of human health is concerned, the environment that exposures due to the release of radioactive material from the repository are negligible compared to natural radiation exposure.

3. It shall be ensured that there is the possibility of retrieving the waste emplaced for the duration of the operating phase and that adequate provisions

are in place for a possible recovery of the waste for 500 years after the planned sealing of the disposal facility.

4. The disposal facility shall be constructed and operated in such a way that no intervention or maintenance is required for the reliable long-term containment of radioactive waste in the post-closure phase.

The preliminary safety analyses (StandAG 2017, §27) in phase one of stage one of the selection process shall assess the extent to which the safe containment of the high-level radioactive waste can be expected by making use of the geological formation given at the respective site. In these analyses, the safety requirements (see above) serve as the normative basis. The Federal Ministry for the Environment (BMU) was authorized by the StandAG (2017, §27) to determine the requirements for conducting the preliminary safety analyses in the site selection procedure by ordinance. This has happened in the meantime. The requirements shall be reviewed every ten years. If necessary, they shall be adapted to the state of the art in science and technology.

The geological exclusion criteria, minimum requirements, and assessment criteria as well as the safety requirements and the requirements for the safety analyses remain valid throughout all three phases of the site selection procedure (stage one of the entire process). From phase one to phase three of the site selection procedure, the criteria will be applied in ever-increasing detail and with ever more accurate data, starting from the use of available (often generic) data in phase one to data gained by surface investigations in phase two and in-depth data from underground investigations in phase 3. This procedure combines the goal of safe and maintenance-free final disposal of the high-level radioactive waste at the end of the entire process with the desire for learning during the process, the retrievability of the waste in case of need, and the reversibility of decisions.

16.3.4 The national citizens' oversight committee

As a result of adverse historical experiences and persistent burdens on the site selection process, the Commission (2016a) proposed a highly ambitious approach with respect to transparency and public participation. The challenge of realizing these ambitions laid down in the StandAG (2017) motivated the Commission to propose an unusual institutional arrangement. In addition to the established structure of an operational and a supervisory institution—in the case of dealing with HLW, this is the combination of BGE and BASE (see above)—it proposed an institutional innovation (Smeddinck 2020). The NBG will ensure the largest coincidence possible between the ambitions of the StandAG and its realization. In the StandAG (2017, §8), it is described as follows (with a slightly different English name than it has today):

(1) The task of the pluralistically composed National Civil Society Board is to accompany the site selection procedure as a mediating and independent body, in particular as regards public participation, with the aim of enabling trust in

> the implementation of the procedure. It may independently and scientifically deal with all questions relating to the site selection procedure, consult the competent institutions at any time and deliver opinions. It may provide to the German *Bundestag* further recommendations on the site selection procedure.
>
> (2) The members shall be given access to all files and documents of the site selection procedure of the [BASE], the [BGE], the Federal Institute for Geosciences and Natural Resources [BGR] and the Geological Services. The results of the deliberations will be published. …
>
> (3) The [NBG] shall consist of 18 members. Twelve members shall be respected public figures. They shall be elected by the German *Bundestag* and by the *Bundesrat* on the basis of an identical election proposal; in addition, six citizens, including two representatives of the younger generation who have previously been nominated in an appropriate procedure of citizen participation, shall be appointed by the [BMU].

In this sense, the NBG is clearly an oversight institution without operational duties in the process of site selection. Instead, it fulfils the roles of monitoring and reflecting the status of the entire process and of the measures taken by the other institutions. In particular, the NBG takes care of public participation and trust building. The significance of this aspect was laid down in the StandAG, which foresaw a dedicated participation officer (StandAG 2017, §8, 5):

> The National Civil Society Board shall appoint a participation officer who, as a member of the secretariat, shall assume the task of early identification of potential conflicts and the development of proposals for their resolution in the site selection procedure.

Apart from the issue of participation, the NBG take cares of learning processes in and during the site selection process. This task involves continuous monitoring (Hocke *et al* 2012) of the ongoing activities and the evaluation of the observations against the ambitions of the StandAG in order to keep the gap between the desired and expected developments, on the one hand, and the factual ones, on the other, as small as possible.

The NBG organizes different types of public and dialogue events. It meets monthly for regular meetings, large parts of which are open to the public and include time for answering public questions (https://www.nationales-begleitgremium.de/). The agenda is published at least one week before the meetings. In order to strengthen the effectivity and efficiency of the NBG's work, working groups (*Fachgruppen*, FG) have been established. They are dedicated to continuous activities and aspects of the NBG: public participation (FG I), geological data (FG II), radiation protection and safety (FG III), and the self-learning system (FG IV). In addition, ad-hoc groups are mandated for dedicated and more short-range purposes.

Recently, the NBG has been concentrating on accompanying the current public participation in the interim report on the subareas published by the BGE in

September 2020 (phase one of the site selection process, see above). In particular, the NBG served as a sparring and discussion partner for the *Fachkonferenz Teilgebiete*, which organizes public participation according to the StandAG and shall bundle the results and forward them to the BGE. This *Fachkonferenz* should be self-organizing in order to avoid external influence, which was and still is an onerous task. The NBG offers feedback platforms after each of the meetings of the *Fachkonferenz* and publishes its own diagnoses and recommendations for improvement. The particular challenges caused by the pandemic, the impossibility of real meetings, debates, and dialogues, and the necessity of using online tools raise questions of whether, to what extent, and in what respect virtual meetings can replace real communication. Early experiences have demonstrated that information supply and Q/A sessions can be done well online, but this is much more difficult with real dialogue and a joint search for common positions. Because of the low effort required for attendance—opening up an internet link instead of travel—people were able to participate much more easily. These effects will be an issue for further consideration by NBG and other actors.

Currently, the NBG is also concerned with preparing for the time after phase one is finished later this year. In phase two (section 16.3.2), no participation is foreseen by the StandAG, which is regarded by the NBG as a serious problem for trust building. According to the interim BGE report (BGE 2020), still 54% of Germany's surface remains under consideration. This must be reduced to a few percent in phase two—and it is hard to imagine that this crucial step can be completed without providing transparency to the public and securing its participation. Furthermore, the NBG is just preparing for organizing an international peer review of the first years of the site selection process and of the results gained so far. Its intention is to learn from external feedback to make possible improvements to the further process and possibly to benefit from experiences in other countries.

16.4 Current status, next steps, and challenges

The relaunch of the site selection process according to the StandAG happened in 2017. The BGE collected all data available from the 16 State Geological Surveys, which are responsible for issues concerning the geological subsurface in Germany. Much of this data was decades old, not always consistent, and not available in digital formats. The BGE expended a huge effort in digitizing this data, feeding it into underground models of Germany and visualizing these models.

A more difficult challenge to the desired transparency was (and still is) the fact that parts of the data fall under private property rights, e.g. from companies that collected data while searching for underground resources. Hence, these data cannot be made public, according to German law. If parts of the data cannot be published, then parts of the conclusions of the BGE concerning the subareas also lack transparency. In order to keep this transparency gap as small as legally possible, the so-called *Geologiedatengesetz* was approved by the *Bundestag* and put into force in summer 2020. To close the remaining gap, the NBG was asked by the *Bundestag* to commission experts who were to check the conclusions according to the NBG's right to access all documents. These experts investigated several cases and approved the methodological correctness of BGE's conclusions in these cases, providing some evidence that the BGE's approach was also correct in other cases. Hence, this procedure could not fully close the transparency gap but could strengthen trust in the BGE's approach.

As indicated already above, the BGE published the interim report on the subareas remaining in the further investigation on time in September 2020. It included the message, surprising to many actors, that the former favorite site at Gorleben should no longer be considered as a possible subarea. The other surprising result was that 54% of Germany's surface should remain in the further process. The BGE explained the approach and the results to the public in an online event in October.

At the time of writing (spring 2021), public consultation and participation responding to the interim report of the BGE is ongoing. According to the StandAG, the *Fachkonferenz Teilgebiete* is organizing this participation, supported by an office provided by the BASE. The *Fachkonferenz* organizes three major dialogue and feedback events. The first, in February 2021, was organized to take place completely online because of the pandemic. This situation raised questions concerning the appropriateness of online communication and motivated the *Fachkonferenz* to postpone the other two events to June and August, hoping that by then at least some real communication would be possible. The results of the *Fachkonferenz* are to be forwarded to the BGE to the end of August. The BGE will, based on this feedback, provide the final interim report to the BASE[3].

Simultaneously, the BGE is preparing for a further narrowing down of the subareas (54% of the landscape of Germany), which will be accomplished by applying preliminary safety analyses (section 16.3.3). This will be a crucial step in the entire process, because a reduction of the remaining areas to a few percent of the German landscape seems necessary to keep the subsequent phase operable. The lack of prescribed participation will be made a major issue by the NBG (section 16.3.4). The determination of the site regions for surface investigation (phase two) will take a few years.

The intensity, complexity, scope, and duration of the repository site selection process make it an exceptional case in Germany. Notably, it uncovered challenges concerning long-term governance in democratic and pluralistic decision-making systems, in particular, in Germany, but also beyond (Kuppler and Hocke 2019). According to the demands and risks, it is not possible to develop a master plan beyond the rough structure provided by the Commission (section 16.3.2). Instead, a robust system of participation that is adaptive and self-healing, in which the real options for participation and the roles in the procedure are clearly defined and transparent for all those involved, is necessary (Kanitz 2020).

Despite the author's personal optimism that Germany has the capability to successfully manage this process in its mature democratic system and culture, the

[3] After the editorial deadline: In summer 2021, after its third meeting, the *Fachkonferenz* submitted its its feedback (https://www.endlagersuche-infoplattform.de/webs/Endlagersuche/DE/Beteiligung/Fachkonferenz/ergebnisse/dokumente-ergebnisse.html). Moreover, ensuing participation formats called *Fachforum Teilgebiete* and *Beratungs- und Planungsgruppe* were initiated in order to address concerns about a lack of participation after the *Fachkonferenz* (cf. last paragraph of section 16.3).

process remains fragile. Besides overarching but abstract threats from possible social or economic instabilities, war or huge natural catastrophes, some concerns are rooted in experience:

- The political consensus, which opened up the window of opportunity after the Fukushima disaster (section 16.2), is fragile by principle, as any political consensus is. The landscape of political parties and their positions undergoes continuous movement. Instead of having a guarantee for the stability of the 2013 consensus, it will remain necessary to renew it again and again.
- The federal system in Germany is often confronted with different perspectives by the states but also by their egoisms. Particular states could run into the temptation to withdraw from the consensus of operating an open-ended site selection process in favor of declaring through top-down decisions that there will be no appropriate site in their own state anyway. In this case, others could also lose their willingness to accept the results of an open-ended procedure.
- In spite of all attempts to take the relaunch of the site selection process and the high ambitions of the StandAG seriously, there is a risk of falling back into old routines at any time. Here, 'falling back' means a return to the routines of top-down government instead of dialogue on an equal footing, calling the police instead of public participation, or demonstrating the power of the state instead of understanding that its power is only ever borrowed from its citizens. This might sound far-fetched in the current situation (2021), but we do not know about future developments. In more than a few democratic countries, populist movements and extreme right-wing parties have successfully occupied public opinion and converted democratic culture within a few years.
- Even in the case of a smooth and peaceful course of the selection process, risks cannot be excluded. A specific risk could be the disappearance of public interest just because of the smoothness and peacefulness of the process. Declining or even vanishing public interest would imply the loss of critical voices from the process. These voices, however, are an important resource for a successful site selection process, because they urge the professional actors to be transparent and to look for learning opportunities, to question themselves, and to avoid the blindness of a closed community. Losing these resources would increase the risk of blindness, of tunnel views, and of missing opportunities for learning and improvement.

It is probable that several other risks could be mentioned. It is important to keep the risks in mind in order to constantly devote a large amount of effort to realizing the high ambitions of the StandAG and to continuously look for improvements.

Bibliography

BGE (Bundesgesellschaft für Endlagerung mbH) 2020 *Zwischenbericht Teilgebiete* https://bge.de/fileadmin/user_upload/Standortsuche/Wesentliche_Unterlagen/Zwischenbericht_Teilgebiete/Zwischenbericht_Teilgebiete_barrierefrei.pdf

Brunnengräber A, Di Nucci M R, Isidoro Losada A M, Mez L and Schreurs M (ed) 2015 *Nuclear Waste Governance. An Int. Comparison* II (Wiesbaden: Springer VS)

Commission (Commission on the storage of high-level radioactive materials) 2016a *Final Report* https://bmu.de/en/download/report-by-the-bundestags-commission-on-the-storage-of-high-level-radioactive-materials/

Commission (Kommission Lagerung hoch radioaktiver Abfallstoffe) 2016b *Verantwortung für die Zukunft—ein faires und transparentes Verfahren für die Auswahl eines nationalen Endlagerstandorts. Abschlussbericht* https://bmu.de/download/bericht-der-kommission-lagerung-hoch-radioaktiver-abfallstoffe/

Grunwald A 2015 Social dimensions of nuclear waste disposal *Key Topics in Deep Geological Disposal: Conf. Report* ed S Fanghänel (Karlsruhe: KIT Scientific Reports) KIT-SR 7696 pp 49–52

Hocke P 2015 Nuclear waste repositories and ethical challenges *Geoethics: Ethical Challenges and Case Studies in Earth Sciences* ed M Wyss (Amsterdam: Elsevier) pp 359–67

Hocke P, Bergmans A and Kuppler S 2012 Gewährleistung von Transparenz bei der Entsorgung nuklearer Abfälle: monitoring als soziale Innovation *Einführung in den Schwerpunkt. Technikfolgenabschätzung. Theorie und Praxis* **21** 5–10

Kanitz S 2020 Safely stored for all eternity. How the Bundesgesellschaft für Endlagerung is conducting its search for a repository for highlevel radioactive waste *atw Internationale Zeitschrift fuer Kernenergie (Int. J. for nuclear power)* **65** 331–4

Kuppler S and Hocke P 2019 The role of long-term planning in nuclear waste governance *J. Risk Res.* **22** 1343–356

Liebscher A, Borkel C, Jendras M *et al* 2020 Towards best possible safety—current regulatory research for the German site selection process for high-level radioactive waste disposal *Adv. Geosci.* **54** 157–63

Radkau J and Hahn L 2013 *Aufstieg und Fall der deutschen Atomwirtschaft* (München: oekom Verlag GmbH)

Smeddinck U 2020 Wie innovativ ist das Standortauswahlgesetz? (How innovative is the Site Selection Act?) *Geschäftsbericht 2019 (Annual report 2019)* Bundesgesellschaft für Endlagerung mbH (BGE) pp 44–7

Sovacool B 2011 *Contesting the Future of Nuclear Power: A Critical Global Assessment of Atomic Energy* (Hackensack, NJ: World Scientific)

StandAG 2017 *Act on the Search for and Selection of a Site for a Disposal Facility for High-Level Radioactive Waste (Site Selection Act—StandAG)* https://bmu.de/en/law/repository-site-selection-act-standag/

www.ingramcontent.com/pod-product-compliance
Ingram Content Group UK Ltd.
Pitfield, Milton Keynes, MK11 3LW, UK
UKHW051858100926
8556IPUK00005B/684

9 780750 330930